PIERCING THE HORIZON

The Story of Visionary NASA Chief Tom Paine

Purdue Studies in Aeronautics and Astronautics

James R. Hansen, Series Editor

PIERCING THE HORIZON

The Story of Visionary NASA Chief Tom Paine

BY SUNNY TSIAO

Purdue University Press / West Lafayette, Indiana

Printed in the United States of America.

Cataloging-in-Publication Data available at the Library of Congress.

Hardback ISBN: 978-1-55753-791-1
ePub ISBN: 978-1-61249-512-5
ePDF ISBN: 978-1-61249-511-8

Cover photos courtesy of NASA.
Jacket design by Lindsey Organ.

In memory of my grandparents, Góng Góng and Pó Pó, and the men and women who courageously fought the Asian Holocaust in the Second World War.

Man's future in space is limitless. We have embarked on a new stage of evolution that will engage all future generations. . . . We must find the answers. We must move vigorously forward in space. The practical benefits alone justify this venture, but there are many other compelling human reasons. Progress in space should continue to spur us onward to find new solutions to our age-old problems here on Spaceship Earth. We must make the blue planet Earth a home base, worthy of men who will set forth one day on journeys to the stars.

—*Thomas Otten Paine*

CONTENTS

Chronology, *ix*
Foreword, *xi*
Prologue: Man Will Conquer Space Soon, *xiii*

1 Navy Brat, *1*

2 I Never Got Over It, *9*

3 A Long Voyage Home, *23*

4 House of Magic, *33*

5 It's a Presidential Appointment?, *45*

6 Gaining Some Respect, *61*

7 I'm Not a Politician, *75*

8 A Great Sense of Triumph, *89*

9 Going Global, *107*

10 What Now?, *119*

11 I Accept Your Resignation, *133*

12 A Little Better Footing, *155*

13 Pioneering the Space Frontier, *167*

14 Chief Martian Monster, *185*

Epilogue: A Twinkle in His Eye, *199*
Author's Note, *209*
Notes, *215*
Bibliography, *243*
Index, *253*

Thomas O. Paine

1921–1992

1920

Born November 9, 1921, Berkeley, CA

Sister Janet is born

Robert Goddard launches the world's first liquid propellant rocket

Charles Lindbergh crosses the Atlantic

General Electric establishes the first television station in Schenectady, NY

Family moves to the East Coast

1930

Japan invades China

Height of the Great Depression

First programmable computer

Hindenburg disaster

Graduates from high school, enters Brown

Germany invades Poland

1940

Japan attacks Pearl Harbor, United States enters World War II

Graduates from Brown, enters the US Navy

Returns from the Pacific, marries Barbara Pearse, enters Stanford

Chuck Yeager breaks the sound barrier

Graduates from Stanford, begins work with GE in Schenectady

1950

Becomes manager of the GE Instrumentation Laboratory in Lynn, MA

Detonation of the first hydrogen bomb

Launch of the USS *Nautilus*, the world's first nuclear-powered submarine

Soviet Union launches *Sputnik 1*

Moves back to the GE Research Laboratory in Schenectady; NASA is formed

X-15 flies to the edge of space

1960

- USS *Pompon* is decommissioned for the last time
- Gagarin orbits the earth; Kennedy commits America to a moon landing
- **Becomes General Manager of GE TEMPO in Santa Barbara, CA**
- US deploys combat units to Vietnam; first manned Gemini flight
- First flight of the Saturn IB launch vehicle
- Apollo 1 fire
- **Appointed Deputy Administrator of NASA by Johnson**
- **Becomes Acting Administrator of NASA**
- Apollo 8 orbits the moon
- **Appointed Administrator of NASA by Nixon**
- Apollo 11 lands on the Moon
- **Resigns from NASA, returns to GE**

1970

- Apollo-Soyuz Test Project; launch of Viking 1 and 2 to Mars
- **Becomes president of the Northrop Corporation**
- Launch of Voyager 1 and 2 deep space probes

1980

- First flight of the Space Shuttle
- **Leaves Northrop; forms Thomas Paine Associates**
- Plans for a space station approved
- **Appointed Chairman of the NCOS by Reagan**
- *Challenger* accident; release of *Pioneering the Space Frontier*
- Bush proposes the Space Exploration Initiative
- Collapse of the Berlin Wall, end of the Iron Curtain and Cold War

1990

- **Appointed to the Committee on the Future of the US Space Program**
- **Dies May 4, 1992, Los Angeles, CA**
- Shuttle docks with *Mir* space station, first US/Russian human flight since 1975
- Launch of the Mars Global Surveyor
- Pioneer 10 passes into interstellar space
- Begin construction of the ISS

2000

- 100th flight of the Space Shuttle
- First space tourist visits the ISS; launch of the 2001 Mars Odyssey
- 9/11
- Barbara Paine passes away
- *Columbia* accident; Shuttle flies for the last time eight years later in 2011

FOREWORD

With the publication of this book, Purdue University Press proudly inaugurates Purdue Studies in Aeronautics and Astronautics, a new family of scholarly books dedicated to the study of flight—both in the atmosphere and in space—in historical, social, technological, political, cultural, and economic contexts.

As readers of the books in our series will learn, the study of aeronautics and astronautics concerns much more than just the nuts and bolts of airplanes and spacecraft. It involves much more than just the history of propellers and wings, more than the history of landing gear and jet engines, more than the ornithology of P-51s and Space Shuttles, or the genealogy of X-planes, rockets, and missiles. The study of aeronautics and astronautics is just as much a story of people and ideas as are studies dealing with any other topic related to society and culture. Without question, scholars who write about aeronautics and astronautics have a lot to say about the research, design, building, flying, maintaining, and utilizing of airplanes, aerospace vehicles, and spacecraft, but their studies are no less human, no less connected to social or political or aesthetic forces, because they deal with technical things. As our books in this new series will demonstrate, an advanced study of aeronautics and astronautics will tell us a great deal about our existence as a thinking, dreaming, planning, aspiring, and playful species.

This first book in the new series is a biography of Thomas O. Paine (b. 1921–d. 1992), one of America's greatest spaceflight visionaries. Not only was Dr. Paine the man who headed the National Aeronautics and Space Administration during the period of the United States' early manned lunar landings in 1969 and 1970, but he also was deeply involved in preparing plans for the post-Apollo era at NASA. We have many biographies and autobiographies of astronauts and many general, administrative, and technological histories of the US space program, but we have too few critical works on the principal managers and bureaucrats responsible for leading and directing the US space program. Fortunately, we now have a close look at the outstanding career of Thomas Paine. Sunny Tsiao offers a penetrating look into Paine's significance as a major figure in the US space program, placing it into the broader context of space history, NASA history, the history of science and technology, American history, and the history of the Space Race.

As with all the publications in our new series, this book should be of interest to a wide group of people, including aerospace scholars, space exploration enthusiasts, those interested in the history of the federal government and federal science and technology planning and management, and the many thousands of people in government, industry, and academe who today are exploring the ways and means of humankind's future in air and space.

JAMES R. HANSEN, PHD
Series Editor
Purdue Studies in Aeronautics and Astronautics
Purdue University Press

PROLOGUE: MAN WILL CONQUER SPACE SOON

Mission Control Center, Houston, Texas—2:15 p.m. Central Time, July 20, 1969

He could see the whole room from where he sat. NASA called it the MOCR, or Mission Operations Control Room, but the rest of the world knew it simply as "Houston," a room born of the space age. The nerve center of America's manned spaceflight program was impressive enough, but was actually quite a bit smaller than it appeared on television. Unless there was a simulation of a spaceflight or an actual mission in progress, Mission Control usually sat empty, with lights dimmed, chairs pushed in under the rows of control consoles, and monitors turned off. Only the whisper of air blowing out of the air-conditioning vents disturbed the silence.

But on this sweltering, humid Sunday afternoon in July 1969, the room was abuzz with pensive excitement. An unmistakable sense of anxiousness, the anticipation of what was about to happen, hung in the air. Mission Control was teeming with flight controllers, mostly young engineers who only three or four years before were studying mathematics and science in college. Now their full attention was on a constant stream of data in the form of numbers and letters that flickered before them on their black-and-white monitors. To the untrained eye, the figures looked like a cryptic alphabet from an obscure, long-lost mathematical language. But to the controllers, the data meant more—much more. And on this occasion, the telemetry had traveled nearly a quarter of a million miles to reach Houston. It was data that was coming from the moon.

From behind a glass wall separating the VIP viewing area from the floor of the MOCR, Tom Paine focused his attention on a greenish-yellow icon on the large projection screen at the front of the room. It slowly made its way across the screen. Shaped somewhat like the odd-looking Apollo lunar module (LM), it showed the position of the faraway spacecraft as it finally began its long-awaited powered descent to the surface of the moon. The final landing sequence would take only twelve minutes, but NASA had been waiting to make that engine burn for eight years.

Voice transmissions coming over the speakers told him what was happening. The voice signals were surprisingly clear, interrupted only on occasion by some garble and static that one would expect, whether listening to a live broadcast of a baseball game from just down the street or, in this case, two men narrating their own landing onto the surface of the moon. Earlier, Flight Director Gene Kranz, the tough, former Saber Jet pilot who was now directing Mission Control's "White Team," had ordered the doors of the MOCR locked. A final status check around the room followed. Each flight controller declared an emphatic "GO!" into his headset.

Two hundred thirty-eight thousand miles away, Neil Armstrong and Buzz Aldrin closed a sequence of circuit breakers that fired lunar module *Eagle*'s descent stage engine to initiate the powered descent sequence. The engine burn slowed them down just enough, gradually taking them out of lunar orbit and onto a predetermined path. At first, only the instruments told them that they were actually descending. But before long, the craters, boulders, and finally the rocks of the moon became very clear. The spacecraft pitched over and the dramatic lunar panorama filled their windows as they approached the landing area. If everything went well, they would be on the surface in the next few minutes.

Four days earlier, Paine was there at the Kennedy Space Center as Apollo 11 left Earth in mankind's first attempt to land on the surface of the moon. The flight was the high point of Project Apollo, America's historic quest to land a man on the moon and bring him safely back to Earth by the end of 1969.

He was in Houston now with the largest contingent of US space officials ever gathered in one place.[1] Only four months earlier, President Richard Nixon had appointed the forty-eight-year-old engineer from California to be the head of the National Aeronautics and Space Administration. On his watch, human beings began to journey away from the confines of planet Earth for the first time on missions to explore another celestial body.

Nearly fifty years later, those historic flights still hold their place as the zenith of America's space program and the accomplishment for which it is most recognized around the globe. From those epochal voyages came paragons of the space age, pioneering and creative geniuses who saw what was possible and made it all happen. Circumstances had now put Paine among those on the brink of making history.

As the descent stage engine continued to burn, the spacecraft slowed, dropping closer and closer to the surface. The icon on the screen showed that the lunar module was on the correct approach trajectory. He was confident that Armstrong and Aldrin would soon be safely on the surface.

For Tom Paine, human beings' exploration of the moon was the first step in pioneering the vast frontier of outer space. In the span of just a few short decades after the turn of the twentieth century, the rapid progress of technology had revolutionized the way people lived. All facets of human society were being advanced to one degree or another. Most tantalizing was that space travel had become a reality.

The years during and after the Second World War had been transformational. Rapid advances in science and engineering made people think that the right use of technology could overcome even the most daunting of society's ills. It was in this dynamic setting that Paine's career began. Computers were still in their infancy; "high-tech" had not yet been invented. He and other engineers of the day creatively devised new ways to apply discoveries in the fields of material science, electronics, and aerospace to everyday life. They used slide rules and vacuum tubes instead of keyboard and mouse. Their ingenuity revolutionized transportation, manufacturing, and weaponry. An ardent futurist, Paine believed that technology held the key to a good future for humanity, not only here on Earth but, one day perhaps, in the borderless expanse of outer space.

Twentieth-century technology of all kinds intrigued him. Ships especially piqued his interest. Neil Armstrong still remembered, just a year before his own passing, that Tom Paine was as fascinated with how airships—those impressively titanic, lighter-than-air, passenger-ferrying marvels of aeronautical engineering that were largely abandoned after the fiery *Hindenburg* disaster—were able to cross the Atlantic as he was with how a fragile spaceship like the Apollo lunar module could fly two men down to soft-land on the surface of the moon.[2]

Second only to space exploration was Paine's lifelong passion for submarines. He made considerable contributions to the field as an engineer. As a historian also, he worked in many ways to preserve the artifacts, histories, and memories of the boats that fought in World War II. The breadth of his work has been recognized in recent years by the US Naval Academy for its unique scholarly and historical significance.

The latter half of the twentieth century saw great leaps and unexpected accomplishments in the field of aeronautics and astronautics. Those achievements became the new barometer by which the prowess of a nation was measured on the world stage. It began in earnest in the years leading up to World War II. Airplanes moved from the realm of mere curiosity to being useful machines for transportation and war.

The pace picked up dramatically after 1945. It reached a crescendo in the 1960s as the competition to be the torchbearer pitted the Soviet Union against the United States. The launch of the world's first satellite, *Sputnik*, in 1957 had made sure of this. President John F. Kennedy wasted no time in responding to an unsure and stunned nation. He challenged America to land a man on the moon and return him safely to Earth before the end of the 1960s.

Sputnik was America's "Pearl Harbor of Space," a wake-up call for a distracted nation. Overnight, the Space Race was born. Disbelief and anger at the Soviet launch consumed America's psyche, from the young who barely understood the meaning of the nebulous images they were seeing on their television screens to the highest leaders at the national level who demanded to know how this could have happened. It was a jarring blow that woke the nation out of complacency and into a technological awakening. It would culminate twelve years later with the flight of Apollo 11.

"A strong demonstration of American technical and military capability is the best assurance we have for maintaining the peace," Paine said after the successful landing.

> I think the space program has had a darn strong influence on what you might call military matters, and I think the demonstration of our capability of landing on the moon was a strong demonstration to the Soviets of the vigor of our leadership and the strength of

> our society that certainly must have given them pause. When the Soviet leaders [ask themselves] the question "How belligerent can we safely get against the United States[?]" [—]I think America's space successes must give them pause when their people say "Well hell, we can do everything the Americans can do and do it better." Well by God, you can't go to the Moon, and I think it has been a stabilizing and sobering force.[3]

Paine saw the accomplishment in still larger terms. He affirmed that while the Moon Race dramatically revealed the difference between two competing, ideologically opposed societal systems and their national values, its true significance was always about how far the United States could push itself. "What the landing on the Moon demonstrates is that American space technology has matured, has come of age. It demonstrates that we can do the thing we set out to do. . . . That's the real meaning of the accomplishment, not that we beat Russia in a Moon Race."[4]

Only ten short months remained to Kennedy's deadline when the US Senate confirmed Paine as the administrator of NASA in March of 1969. The timing of his appointment was important and, for him, incredibly fortuitous. The three highest-profile missions of the moon program (Apollo 8 was the first to orbit the moon; Apollo 11 was the first to land on the moon; and Apollo 13 was the only failed attempt to land on the moon) all flew on his watch. Those epochal missions that sent human beings away to explore another place in outer space have not been attempted since the final Apollo flight splashed down in the Central Pacific in December of 1972.

The timing of the lunar program was strikingly compelling. The 1960s was a restless time in America. Social activism, generational distrust, the civil rights movement, and the Vietnam War defined the decade. A weary public openly questioned what the country was doing in space when the cities were in chaos and young men and women were dying in the faraway jungles of Southeast Asia. The young, especially, questioned authority, the government, and what the American way of life stood for.

Going to the moon became a way to inspire, challenge, and bring the contentious generations together. It suddenly became very important to the nation's psyche and sense of security that the United States reach the moon first. Paine believed that the space program could accomplish something special, something unique for America and even for the human race. The moon landings proved that a country working together could do something remarkable. As World War II had shown, a challenged nation could bring great power and wealth to transform the world—NASA did the same.[5]

Paine was able to influence the space program at a very high level, and he certainly wanted it all when it came to space. National leaders listened to him. Critics who found him too ambitious and his ideas too untenable listened nevertheless. He made his mark as the administrator of NASA when mankind went to the moon, but his articulate championing of a strong human presence in space in the final decades of the twentieth century may, upon revisiting, turn out to be his greatest contribution and legacy.

In the mid-1980s, the Reagan White House asked Paine what America should do next in space. He responded by championing a plan for how human beings, in the span of one generation, could settle the inner solar system. After the pioneering flights of Project Apollo had faded into the pages of history, he remained vigilant and called for the United States to pick up where it had left off: return to the moon and go on to Mars. Now, one score and five years after his passing, those decisions and imprints can still steer the space program as the US considers charting a way back into outer space beyond the horizon of low-Earth orbit.

A renowned fellow presidential commissioner fondly recalled Paine as a man who was "a wonderful human being who was very shrewd but never gave the impression [of] being shrewd. He . . . came across as being just kind of an ordinary guy, but he was quite extraordinary."[6] To see how Tom Paine came to be a central figure in the US space program, we must open not with the moon and Project Apollo, nor the romance of the high plains of Mars that he hoped human beings would one day settle, nor the saga onboard a US submarine as it perilously fought the Japanese, but in the colonies of the New World at the time of the birth of America. This was a time when visionaries of another kind journeyed across the breadth of an ocean to conquer their dreams.

1

NAVY BRAT

If there must be trouble, let it be in my day, that my child may have peace.
—Thomas Paine, 1776

Freetown, Massachusetts, is on record as being one of the very first settlements in the Plymouth Colony and one of the earliest towns in the New World of North America. As one of the first parishes in the New England territory, it was home to some of America's earliest patriots. Most were direct descendants of the Pilgrims who had arrived on the forested banks of the winding Assonet River around the year 1660. There, they began building new lives, befriended the local Wampanoag Indians, and bought parcels of land from them. The settlers engaged in the trade of furs and textiles and cultivated the rich agricultural resources of the area, producing an abundance of grist, in particular, that they profited from by selling to neighboring communities. By 1685, the township had grown large enough that it was incorporated into the governance of the much larger community of the Massachusetts Bay Colony.

An early town census recorded a grand juryman by the name of Ralph Pain living in the township around the turn of the seventeenth century, when the area was growing rapidly in population and importance. It mentions that he served at one time as a constable of Bristol County, Massachusetts. His lineage would remain in the pastoral community through the better part of the century leading up to the American Revolutionary War.[1]

Before 1776, the quiet settlement was known as a Tory stronghold, friendly to loyalists of the British Crown. "God, King, and Country" controlled the politics of the town hall and the rules of the parish. Despite this, by that year, quite a number of the townspeople had become well engaged in the rapidly growing separatist movement. There is evidence to suggest that Job Paine, one of Ralph Pain's grandsons and a direct ancestor of Tom Paine, was proscribed by townsmen as a well-known Tory.[2] On May 25, 1778, a British ship had sailed conspicuously into lower Freetown, its true intentions still a topic of debate to this day. Most historians are of the opinion that it was done to openly provoke the separatists. If so, they did not have to wait long, as a skirmish broke out when a few local minutemen opened fire. Some one dozen commoners armed with muskets then fought off over 150 British marines before the ship retreated a few days later to loud cheers of "Huzza!" from the victors.

In the same well-established Plymouth Colony was one George Soule, an indentured servant who came to the New World on the *Mayflower*. Soule would go on to sign the Mayflower Compact, the first governing document of the colony. According to Tom Paine's grandfather, their family ancestry included Soule, along with *the* Thomas Paine—the consequential pamphleteer of the American War of Independence.[3]

For a century and a half, from about 1692 to 1847, four generations of Paine's family toiled, married, and died in Massachusetts. From there, the family picked up and moved south to nearby Rhode Island. There, his paternal great-grandfather settled and put down roots for the family. In Providence, his grandfather, Frank Eugene Paine, had some influence for a while as a popular state senator. In 1893, his wife Jemima bore their second child in the town of Warwick, the second largest community in the state, and by chance the site of the first shots fired in the Revolutionary War.

George Thomas Paine, father to Thomas Otten Paine, was known to be a high achiever from an early age. He studied civil engineering at nearby Brown University. Nearly half a century earlier, in the year 1847, the seventh oldest college in the United States had broken new ground by instituting one of the first comprehensive engineering curricula in the country (the first to do so was the US Military Academy at West Point). After graduating from Brown, he went on to study naval architecture, first at the Massachusetts Institute of Technology and then at Harvard University. After completing his studies, he joined the US Navy Construction Corps. There, he made his mark by specializing in the engineering and operation of submarines.

The Navy commissioned him a lieutenant (junior grade) in 1917 and sent him packing west across the country to the new base at San Francisco Bay. That was where the US had established its Pacific Submarine Fleet in June of that year. Only two months earlier, the country had entered the First World War following three years of diplomatic neutrality. Paine steadily rose through the ranks, rotating assignments at various stateside naval bases every few years. His wife, the former Ada Louise Otten, gave birth to their first child, Thomas Otten Paine, on November 9, 1921, in nearby Berkeley. Two years later, their second child, Janet Augusta Paine, was born in Arlington, Virginia.

After a distinguished thirty-year career in the Navy, George Paine would retire with the rank of commodore in the years following World War II. He would go on to be a director of American President Lines. There, he oversaw the operation of the company's luxury steamers in the final years of the golden age of the ocean liners that crisscrossed the Pacific and the Atlantic before the advent of the jet age.[4]

Tom Paine was a typical "Navy brat." He and his sister lived wherever their father's duties took him. For the most part, this meant moving to a different base on either coast of the country every few years. His first three years of grade school were spent in Washington, DC. When he was eight, they moved back to the Bay Area, where he finished grade school in Vallejo, California, near the Mare Island Naval Shipyard.

Tough times had hit America in the 1930s. The Great Depression had sent the US and other Western countries into a deep economic tailspin. Poverty and unemployment afflicted the families of many of Paine's schoolmates and friends. The stock market crash and the financial breakdown that followed had turned the American dream into the long nightmare of a dustbowl for a whole generation of hardworking families. Being a Navy officer, George Paine was able to shield his family somewhat from the widespread hardship. Even with the world economy in the abyss, the US Navy was still building up its fleet and providing jobs to long lines of waiting workers.

Local shipyards, especially, looked for able boys and young men to work all kinds of odd jobs. During the summers, Paine worked as a welder in the yards around the harbor. The experience was good for the young teenager. As an apprentice shipfitter, he worked with men of all ages repairing many old ships in dry dock. From this he learned the finer points of shipbuilding and good workmanship. Long days working in the squalid conditions of the yards taught him firsthand the nuances of teamwork and the importance of getting along with those on whom he depended to get the job done.

Ocean vessels, and in particular, submarines, became his passion. Young Thomas was absolutely captivated. His father had given him a model S-boat when he was just five. (He would keep the model for six decades.) George Paine often took his son with him to the shipyard, where he would stay all day. He learned the sailor's trade and mingled with machinists' mates, deckhands, and stokers, gruff workers whose hands were rough and fingernails dirty. Peering into many periscopes as a Navy Junior soon sold him on the high adventures of one born to prowl the depths of the sea.[5]

After three years in California, the family moved back to Norfolk, Virginia. George Paine pushed his son academically. Math and science took top priority. Tom performed well enough scholastically throughout high school and graduated near the top of Matthew Fontaine Maury High School's Class of 1938.[6]

Military pomp and circumstance, parades, firing of cannons, and ship christenings were all common year-round activities for him. George Paine was a career military man. From a young age, he instilled in his son a strong sense of patriotism rooted in traditional American values. Those who knew him described Tom as a personable yet not overly gregarious boy. He developed some uncommon pastimes that require one to pay close attention to details to receive their full enjoyment: book collecting, sailing, and beachcombing—leisure activities he carried into adulthood.[7]

His interest in sailing and seafaring had already been cemented by age ten. That was when his father gave him a first edition of the book *How to Build the Racing Catboat Lark* by the Rudder Publishing Company. He recalled that he easily devoured the pages even at that age. Using the simple drawings in the book, he spent a summer building a seaworthy sailboat and tested it in the currents of the Napa River by the San Francisco Bay.

In Norfolk he again used it, inviting his high school friends to brave the eddies of the Chesapeake Bay with him.[8]

In the summer of 1938, he had a decision to make. Coming out of high school, the one and only place he wanted to go was the nearby US Naval Academy. He wanted to follow his father's footsteps into the Navy. Having been raised and rooted in the values of military tradition his whole life, that had been his plan all through high school.

His application was proceeding smoothly until his physical screening. It was then that the medical examiner told him the bad news. While his eyesight was not poor enough to keep him out, it would probably worsen after four years of studies to the point that he might not qualify for an officer's commission. He would not be going to Annapolis, although not by his own choice.

Tom Paine was devastated. But he still wanted to study engineering. The Massachusetts Institute of Technology was not too far away, so he turned his attention to the country's most prestigious engineering school. But George Paine thought his son, not yet seventeen, would be a bit young for the academic rigors that he would encounter there. He advised Tom to instead try his own alma mater, Brown University. There, he could get a more diverse education in the liberal arts as well as exposure to a curriculum in engineering. He could then try to go on to graduate school at MIT. Tom had not seriously considered Brown up to that point, but he followed his father's advice.[9]

In the fall of 1938, the sixteen-year-old left home and went up the coast to Providence. Of the 470 incoming freshmen that year, he was among the youngest. He began a general education program in the arts and sciences but with an eye toward majoring in engineering. Classes in calculus, European history, creative writing, mechanics, and classics were all required. He later reflected quite candidly that as a student he was probably just a "little better than average."[10] It showed on his transcript. He would receive an A if the subject interested him, but failed a semester of German because he did not see the logic in memorizing vocabulary. In his third year, he refined his studies to concentrate more heavily on the discipline of electrical engineering.

Initially taught as a discipline of physics, the study of electricity as an independent field of engineering matured rapidly in the first decades of the twentieth century. Since the late 1800s, electricity had dramatically transformed the way people lived. By the 1930s, the field was an established flagship of modern engineering sciences, taught at universities around the world along with chemical, civil, industrial, mechanical, and later, aeronautical engineering. Demand for electricity was growing rapidly. The widespread, everyday use of industrial and home appliances, long-distance power transmission, lights, and radios, had made it into the high-tech industry of its day.

Tom pledged to a fraternity, as was common for all gentlemen of that era. The school's chapter of Delta Kappa Epsilon elected him over three other candidates to be the vice president. Founded in 1844 by fifteen Yale students, the East Coast fraternity, while not exclusive, was certainly prestigious, having long had a well-established reputation as a rather elite society. Members of the national organization included five presidents and three Supreme Court justices. Outside of class, he worked for the Brown *Daily Herald* newspaper and the university's John Hay Library. To relax and socialize, he pursued his favorite pastime: sailing. He joined the school's yacht club and spent many warm weekends sailing the Narragansett Bay and the sites around Warwick and Bristol. On occasion, the club would test their skills out past Nantucket Sound in the scenic surroundings of Cape Cod Bay. Tom's four years of college were quite ordinary, by all accounts. In May 1942, George and Ada watched their son receive his bachelor's degree in engineering.[11]

He was ready for MIT academically and had been preparing for graduate studies there while he was still at Brown. The country, along with most of the other industrialized nations, was finally beginning to work its way out of the mire of the great global economic depression. But the US entry into World War II had changed his plans. The rest of the world had already been at war for several years. In Asia, decades of skirmishes had exploded into a full-scale, all-out conflict after the Empire of Japan invaded the Republic of China in 1937. Two years later, the German Third Reich plunged Europe into apocalyptic war with a blitzkrieg attack on neighboring Poland. The United States assisted China and Great Britain with supplies and volunteers, all the while trying to stay out of the conflict. But December 7, 1941, changed everything, as war came unannounced to the country's doorstep.

2

I NEVER GOT OVER IT

We're going out to sink enemy ships, but we are coming back.
—Stephen H. Gimber, Commanding Officer of the USS Pompon

The *Brown Daily Herald* was just about to go to press when the news hit. He took his coat off and sat back down in the university newspaper office. They needed to stay late and get the paper's first extra edition out as soon as they could. He glanced at the clock. It was almost five in the afternoon on the East Coast. In Hawaii, it was not yet noon. The attack had already been over for two hours. By now, the Associated Press wire service out of Washington had confirmed the news.

December 7, 1941, changed people's lives. The morning after Japan's sneak attack on Pearl Harbor, Tom and a group of classmates from Brown took the short train ride to Boston. Classes were not very important that day. They had something more pressing to do. At the War Department recruiting office, most signed up with the Army Reserve. He and his best friend from the yacht club applied for admission to the US Naval Reserve Officers Training Program.

Paine was just finishing up the first semester of his senior year. His intention to serve in the Navy had not diminished since the time he was dissuaded from applying to the Naval Academy due to his poor eyesight, and America was now at war. The Navy had no problem with him this time around and quickly processed his application. On January 8, 1942, he was notified that he had been accepted as a Class V-7 Seaman Apprentice in the reserve midshipman program. The armed forces needed numbers quickly, and this was one way, a very common way, for college graduates or men who were about to complete college to enter wartime service as apprentices.[1]

In a twist of irony, the Navy assigned him to Annapolis. After graduating from Brown, he reported on September 11 for three months of intensive basic training at the academy. He was one of dozens enrolled in an accelerated program that would give him a Navy commission in just a few months (the so-called "90-day wonder").

Eighteen months of curriculum were packed into three short months. Classes in basic naval indoctrination, fitness training, and military tradition were required. There was also chapel at noon every Thursday. After a month, he was commissioned as a midshipman for Engineering Officer Training. His class notes on electrical engineering, thermodynamics, and navigation were neat and exact. Radio communications that stressed "accuracy, security, [and] speed" he found especially engaging. His natural penchant for mathematics gave him good marks at the academy. He had been sailing since he was ten, and coursework in ocean navigation came easily to him. It was all not too different from the studies he had completed only three months earlier at Brown. Two months later, he graduated from midshipman training. Paine was now an ensign, E-V(g), in the US Naval Reserve.[2]

With a Bachelor's degree in engineering, the Navy appointed him to the line of general engineering services. He accepted this, but there was only one assignment he wanted. For as long as he could remember, he had been deeply fascinated by submarines. Growing up near the San Francisco Bay had fostered that desire. There, he had watched with wonderment as submarines maneuvered across the Napa River from the Mare Island Naval Shipyard. On some days, he would watch for hours. It had stirred his imagination of life in the steely beasts of the deep. Through his spectacles, the boy's marveling blue eyes would patiently survey the choppy surface of the water for any hint of a slow-moving periscope or ripples from a subtle wake.

The first use of underwater, iron war vessels in modern times was in the American Civil War. Both the North and South had experimented with crude, man-driven, leaky submersibles that they used to ram unsuspecting ships with explosives at night. Almost all became lonely coffins of steel.

Revolutionary advances took place over the next fifty years. Diesel-electric boats had become a staple of the modern war arsenal by the turn of the century. They had a deadly impact during the Great War; victims of the German U-boats numbered in the thousands. With the lessons of the war, the technology of submersibles grew and advanced quickly. By the 1920s, the United States had moved steadily to match and then exceed the capabilities of the U-boat.

The Navy granted Tom's request for submarine duty. Ten days after completing the Reserve Officers' Training Corps (ROTC) program, he reported for service. His first assignment was as a student officer at Submarine Division 12 in Key West, Florida. The USS R-14, a small, rather unimpressive old boat, as he recalled, became his first submarine. The R-class had entered service just after World War I, but had been used only to train new recruits since 1930. (The R-14 was, coincidentally, one of the "old rust buckets" that the elder Paine had been stationed on in 1921, doing deep submergence testing off the coast of San Francisco.) It performed "clockwork mouse" duty, repeatedly diving and surfacing all day long in the warm, calm waters off south Florida. There, he got his first taste of how to run a submarine.

During a refit of the R-14, he transferred to the R-10. It, too, was a World War I–era vessel that needed a lot of work. For nearly three months, he learned the ropes of surface and submerged boat handling and underwater navigation. He confided in his journal of his anxiousness one day in the first tense moments upon seeing a spurting rivet pop like a firecracker just feet away, and the feel beneath the soles of his shoes of the hull creaking as if about to split open on his first dive to 200 feet below the surface. He learned and relearned the intricacies of operating the Kingston engine valves and recognized the unmistakable shrill of the balky air injection engines. That basic knowledge would one day save his life in the Pacific.

From Florida, he headed north to the Atlantic Naval Submarine Base at New London, Connecticut. It was home to the country's vast Atlantic Fleet. When he reported for duty on March 29, 1943, it was his last stop before deploying to the Pacific. At New London, he received his qualification in the next class of boats, the S-class. It was a much more difficult class to operate than the R-class. The S or "Sugar" boats were larger and had much greater dive endurance for a more realistic simulation of combat conditions. The three months of intensive training off Long Island Sound came and went quickly. His next stop: the war in the Pacific.[3]

The tide started to turn by 1943. When Japan invaded Manchuria in July of 1937, it had triggered the Second Sino-Japanese War. This led to the attack, four years later, on Pearl Harbor and America's entry into World War II. Encountering no effective opposition at the beginning, the powerful Imperial Japanese Navy (IJN) expanded quickly eastward. By the summer of 1942, it controlled nearly the entire Pacific west of Hawaii. Then things began to change. Allied forces scored strategic victories at the Battle of the Coral Sea and at Midway. The momentum continued in the hard-fought and drawn-out Solomons Campaign throughout 1943.

Following Pearl Harbor, the United States initiated unrestricted submarine warfare against Japan and Germany. Warships and merchant vessels were attacked on both sides. There were no warnings and no aid was given to survivors. Submarines were less than 2 percent of the US Navy's capability, but they inflicted over half of Japan's merchant marine losses during the war. Teaming with forces of the British Navy, the US-led Allies seized control of Japan's supply lines and deployed a sprawling naval blockade that slowly starved the imperial war machine of badly needed natural resources. Chief among them were iron ore and crude oil. Submarine bases were established in Cavite, Fremantle, Brisbane, Sri Lanka, Ceylon, and on Midway Island.

On June 26, 1943, Tom headed out to the Pacific. With a standard-issue duffel bag heavy on his shoulder, he boarded a train to San Francisco. From there, he caught a ride on a B-24 Liberator that was headed to Oahu. Twelve hours later, he saw the sobering remains of a "sunken Battleship Row in oily Pearl Harbor," as he later wrote. Sitting tall in his jump seat looking out a side-gun window, he saw Hickam Field all "shot up with many bullet holes and bomb cavities." In an instant, it brought home to him the solemn reality of war. From Oahu, he continued on, island hopping from Christmas Island to Samoa, Fiji, and finally, Brisbane. His final stop was on the other side of Australia, in Fremantle. As they approached the harbor, he saw a fleet of boats neatly lined up along the Swan River, charging their batteries. He recalled that it was a "most memorable, thrilling sight!"[4]

The Fremantle Submarine Base was the second largest US submarine base in the Pacific. It was a busy place in the summer of 1943, serving as a hub for logistics, boat repair, and command post operations. Located just south of the city of Perth, it was the base of operations for the entire Allied submarine effort in the Southwest Pacific. Nestled in a well-protected rear area away from the frontlines on the southwest coast of Western Australia, Fremantle was safely isolated from any direct Japanese aggression. Upon arrival, he reported to the Seventh Fleet Commander of Submarines, Southwest Pacific Force, but was told to stand by; dozens of newcomers were arriving every day. A duty officer told him that he could find a bunk on the submarine tender USS *Pelias* (AS-14) and put Paine's name down on the duty roster under the commander of Task Force 71, Submarinc Squadron 6.

The fleet was short on officers when Paine was spotted one day by a logistics chief who, on the spot, made him the relief crew officer for Submarine Squadron (SUBRON) 6 and SUBRON 16. His job was to oversee the refit of boats as they returned from sea. He had to make sure that each vessel headed back out to sea was fully stocked with food, supplies, and ammunition. Wartime provisions were high-value commodities, and he kept detailed records of everything that he ordered put onto a boat. During this time, he saw the USS *Billfish* (SS-286) out to sea and the USS *Tuna* (SS-203) set sail on her ninth war patrol.[5]

On November 15, 1943, five months after arriving in the Pacific, he received orders to report to Lieutenant Commander Earle C. Hawk aboard the USS *Pompon* (SS-267). Meeting in a small makeshift office by the barracks, Hawk looked at his file, asked him a few questions, and then made him his assistant engineering officer. The "*Peaceful P*" had just come back from her second patrol of the war and some of its officers were being reassigned. Hawk also put him in charge of restocking the boat before she headed back out for her third patrol. Since he had taken a radar repair course when he first arrived in Fremantle, Hawk also made him his radar officer.

Boats of the Gato class were the largest submarines in the Allied fleet. At 311 feet, they were as long as a football field. The size was necessary owing to the extremely large patrol areas in the Pacific. A typical patrol could last anywhere from sixty to seventy-five days at sea. The backbone of the American fleet, seventy-seven were built, of which twenty were lost. Like most fleet submarines of the time, the *Pompon* was named after a marine creature. In her case, it was a small fish found in the brackish waters of the Louisiana bayous. Commissioned on March 17, 1943 at the Manitowoc Shipbuilding Company in Wisconsin, she went on to earn four battle stars in nine war patrols, with Paine sailing on the last seven.[6]

Paine found himself on the bridge on the afternoon of November 29 as the *Pompon*, with her crew of sixty, quietly slipped out of Fremantle for the shipping lanes of the South China Sea. They had a secret mission: emplace eleven Mark 12 magnetic mines in a corridor near the shallow waters of Poulo Condore off the French Indochina coast. The newly modified mines had just arrived in the Pacific. The *Pompon* was carrying the latest high-explosive technology of the war.

Hawk had briefed his officers on their mission. Naval Intelligence had homed in on the high-value, narrow corridor weeks earlier. On paper, the US strategy was simple: force as many Japanese vessels as possible out into the deeper waters of the channel using the mines. Once in deep waters, they became vulnerable to Allied submarines. The IJN would then have to divert more minesweepers and escorts into the channel. This would leave shipping operations even more unprotected and exposed in other parts of the open waters around Japan.[7]

On board with him was an experienced and close-knit group that included Bill Mendenhall, Frank Wall, Ben Franklin, and Carl (Army) Armstrong. A young lieutenant, Walter H. F. (Wally) Wahlin, an experienced officer who had been on the previous patrols, became his tutor. Once at sea, Hawk made Paine the boat's junior officer of the deck under the watch of Wahlin. Wahlin took him under his wings.

"Suppose a ship appeared suddenly about 15 degrees on the port bow and it was heading right for you, what would you do?" Wahlin quizzed him on his first watch out of Fremantle. Paine thought he had a good answer. "Ring up flank speed and put the rudder left full; head right for him so if we collided it would be our bow in him. I'd sound the general alarm and the collision alarm too, and have the lookouts man the after 20 millimeter gun to clear his bridge and deck as we passed close aboard."

It was not a bad answer. Still, Wahlin corrected him. "You'd be all right if you did that, but don't sound the collision alarm; the crew can't get to battle stations if you've sealed the ship up for collision. . . . If you rammed him, your after 20 millimeter wouldn't bear, so you'd better get the quartermaster up to man the after gun and have the lookouts open fire immediately with the forward 20 millimeter on his bridge thus confusing his helmsman and making it harder for him to ram."[8] He had found a teacher and friend in Wahlin. He could trust Wally.

He wrote down his new experiences in a wartime journal: ocean refueling in Exmouth Gulf from a torpedoed Dutch tanker; the daunting forty-foot waves of Darwin Harbor; night watch on the pitch-black waters of Lombok Strait, deep in enemy territory; the first crash dive at dawn in the Java Sea; the bone-crushing violence of depth-charging in the Palawan Passage that felt like sledgehammers hitting an immovable object. Other entries were hypnotically blissful: the picture-perfect azure waters and lush mountains of Borneo and Celebes; the tropical beauty of Bali; the majestic volcano of Mount Agung; the silent, moonless night of Lombok Strait.[9]

They had been patrolling the shipping lanes off the southern coast of Japan for weeks and had not seen a thing. Unlike the German U-boats that hunted the North Atlantic in packs, described by Winston Churchill as a "measureless peril," US submarines in World War II largely patrolled alone. Like snipers in the sea, their strategy was to lie in wait for weeks for an unsuspecting vessel to come along. By May 1944, the *Pompon* had silently made her way deep into the enemy waters of the Kii Channel and Bungo Strait, the north and south outlets from the Inland Sea that directly controlled Japan's access into the open waters of the Pacific. Naval intelligence had given this tactically important area the code name Cello in the greater Japanese coastal area code-named Hit Parade. They were in Japan's backyard. Such close inshore support usually meant trouble.

On May 30, his submarine made contact with a lone vessel steaming slowly just off Muroto Zaki, a prominent point on the coast of southeast Japan. Lieutenant Commander Stephen Gimber, who had taken over for Commander Hawk, sounded general quarters.[10] Paine was on the bridge for the first time as the diving and torpedo-gunnery officer.

A clear sonar contact had been made at 8:32 in the morning. For the next twenty-four minutes, they quietly stalked an unsuspecting vessel and slowly maneuvered into firing position. It was a two-thousand-ton troop transport (later identified as the *Shiga Maru*) that was steaming home from the East China Sea.

By 8:56, they had made their way directly broadside of the convoy and were in good position to take a shot. But they nearly lost their chance. Only two minutes earlier, with no warning, the *Pompon* had suddenly lost power. Paine's quick action on the bow planes stabilized the nose and brought the boat back into the correct firing position. With crewmen Paul Stolpman and Whitey Bevill working the forward torpedoes, he was able to confirm the forward tubes ready just in time. Gimber immediately gave orders to fire. Three torpedoes shot out from the bow tubes toward the convoy 2,300 yards out. Ninety seconds later, the transport ship was gone. Gimber wrote in the mission log: "Bulls eye! One hit amid ship and he literally disintegrated, breaking in half and sinking almost immediately. Numerous breaking up noises were heard; in addition, two other explosions which were probably the other torpedoes on the beach." After the war, Japanese shipping logs revealed that the captain of this vessel had requested to go around the western side of Japan because he believed that there was too much danger from US submarines on the Pacific side. His request had been denied.[11]

After confirming the kill, Gimber had Paine quickly put the boat into a steep dive and level out at a depth of 100 feet. This standard tactic was usually enough to get the submarine out of immediate danger. But to their bewilderment, a Japanese aircraft quickly found them. Three surface ships zeroed in; two more soon joined the hunt and boxed in the *Pompon*. The team methodically worked the submarine over, helped in part by the *Pompon*'s own intermittently noisy portside propeller. The Japanese dropped some sixty depth charges over the next eight hours. Gimber had to use all of his experience to keep his men alive. He repeatedly dove the boat deeper and deeper to evade the charges that rattled the men and the machine. The crew listened for the dreaded sound of water bursting through the superstructure as they sweated it out for the next nine and a half hours, life preservers ready. The deadly game of cat-and-mouse continued all day and through nightfall. The hunters finally gave up when the *Pompon* was able to play dead. It then quietly slipped away with its batteries nearly drained as nighttime befell the waters.

Naval Intelligence later revealed that the very effective pursuit by the Japanese was one of the first uses in the war of a new magnetic airborne detector. Called the Jikitanchiki, it could locate a submarine hiding as deep as five hundred feet. Paine's boat nearly became its first victim. The first depth charge dropped by the aircraft had pinned the sub down as ships on the surface took up positions to box her in. The precise operation was almost successful. They were able to survive partly because sonar conditions happened to be poor that day.[12]

Weeks then passed without any hint of danger. They spent many long hours waiting for the enemy. Paine found that the high anxiety of encountering the enemy could change, in a moment's time, to the exhilarating feeling of freely sailing the open seas. Between the times of intense action punctuated by sheer terror were long periods of calm and restful serenity.

At the communications console, he would tune the boat's radio to the BBC Overseas Far Eastern broadcast. There, he searched the dials to find Dame Vera Lynn poignantly singing "There'll be Bluebirds over the White Cliffs of Dover." Turning on the intercom, the alluring, strong voice of "The Force's Sweetheart" would echo through the bowels of their steely home-at-sea. He would stop what he was doing, close his eyes, and listen to old favorites like "We'll Meet Again" play over and over. "Not only did it relieve the boredom, it gave us something to hope for," he reflected long after the war.[13]

On August 12, things changed in a hurry. The *Pompon* had made its way north just off Sakhalin Island in the Sea of Okhotsk. Lookouts had spotted three Japanese ships and their escorts earlier in the day steaming southward along the coast. With visibility very poor, Gimber had decided to wait for the cover of darkness to launch a more risky, but potentially more effective, close-range surface night attack.

By evening the sea was glassy and calm. Shortly after nine they broke the surface and quietly approached the convoy. Paine began softly calling out the bow angles at 6,000 yards. At twenty minutes to midnight, Gimber ordered three torpedoes out of their tubes in rapid succession. Two lit up the night, striking the number one ship, an 8,000-ton oil tanker that was soon on fire. She turned hard to port to launch another spread of three torpedoes at the number three ship, a 4,000-ton cargo transport later identified as the *Mayachi Maru*. It sustained two hits, broke in half, and went to the bottom.

Paine heard a loud, sharp slap at the bow and knew exactly what had happened. One of the torpedoes had gone erratic, made a circular run, and acquired the *Pompon* as its target. There was not much they could do but keep track of their relative positions. "For the next two or three minutes, I was extremely busy with our own noble experiment in a nip and tuck race until it passed our starboard quarter on a 30 degree track well inside of 200 yards."[14]

The torpedo meandered unpredictably about the boat like a 3,300-pound blind fish and disappeared. This kind of frightening calamity was not all that uncommon. The US had problems with its torpedoes throughout the war, most notably with the Mark 14 steam turbine model that was standard issue on the fleet submarines. Its technology lagged far behind the Japanese during the war. Paine would later tell *New York Times* journalist Thomas Buckley that being sunk by his boat's own torpedo would have been "most disappointing."[15]

The submarine had, by then, been cruising for some time with a leaky sea valve on her number three sanitation tank. Gimber was hoping that repairs could wait until they reached base back at Fremantle or Midway. But the leak worsened and they could not wait any longer. Given an all clear at the stroke of midnight on August 10, Paine went over the side. Earlier in the day, he had raised his hand when Gimber asked for a volunteer. Wearing a diving suit, mask, and air hose with light only from a bulky waterproof flashlight, he dove deep to seal the valve by hand.

The Sea of Okhotsk was twenty-nine degrees Fahrenheit, foreboding, and numbing. Underneath the steel hull was total darkness. Twenty minutes passed, then thirty. Without word, the rest of the crew waited nervously, their submarine unable to dive as Paine finished sealing the valve. Forty minutes later, he came up to the surface, exhausted but giving a "thumbs-up." For this meritorious action, he would receive the Commendation Ribbon from the commander in chief of the Pacific, Admiral Chester W. Nimitz. Gimber also recommended him as "Qualified in Submarines" to the chief of naval personnel. The qualification was granted by the commander of Submarine Division 43 and entered into his service record on September 25, 1944.[16]

To finish out its seventh war patrol in January 1945, the *Pompon* made its way to the shipping lanes of the Yellow Sea. On watch one morning as the gunnery officer, he noticed an escort aircraft acting strangely off the boat's port bow. As the "friendly" got closer, he noticed the sudden gleam of "red meatballs" on the wings of the aircraft as it started its dive. What he thought was an escort was actually a lone Japanese fighter (a Nakajima Ki-43 Oscar) patrolling the coastline. After a couple of passes that left machine-gun slugs spattered across the conning tower, it flew off into the clouds. But their haste to escape the surprise attack caused the boat to go into a dive with a conning tower hatch stuck open. Seawater poured in and the boat was heavily flooded. Several men struggled to secure the faulty hatch. Listing heavily to one side, the *Pompon* limped to Midway Island thirteen days later, its deck almost awash. The base commander was astounded. As they stood on the dock talking, he told Paine that it was probably the most damaged submarine that had ever made it back to Midway.[17] Of the harrowing escape, Paine would write, "The vague realization that it would be a long, long war, with death never very far away gripped my imagination. Here we were, this was all for real; we were swept up in incalculable violence."[18]

On July 21, 1945, they rendezvoused with the destroyer USS *Herndon* (DD-638) and headed to Guam to close out the ninth—and what would turn out to be the final—patrol of the war. Not knowing when the war would be over, Paine and the others still clung to the hope that they might return home to San Francisco and see "the Golden Gate in '48." By July, however, an unmistakable quietness had swept across the sea. Japanese shipping had dwindled and soon stopped entirely. Marine forces had completed their costly island-hopping campaign toward the Japanese homeland. Preparations for Operation Downfall, the invasion of the Japanese homeland planned for the spring of 1946, were already well underway by then. The relentless bombing of the once sacred and untouchable Japanese soil by B-29s continued—then suddenly stopped, followed by word that the Empire of Japan had surrendered unconditionally.

The fighting may have been over, but for Paine, more exploits and a historical encounter awaited him in the Pacific. Three weeks before General Douglas MacArthur accepted the surrender of Japan on the deck of the USS *Missouri* in Tokyo Harbor, Paine received orders discharging him from the his duties on the *Pompon*. The Navy had him stay behind on Guam, however. His job was to record the time, place, and circumstance of every US submarine sunk by the Japanese. The paperwork was all quite necessary. There was still much to do to clean up after the war.

In the days following, he began sifting through the Japanese records with the help of a translator. After reviewing just a few files, he concluded that the written records were of very little use. Throughout the war, the IJN had greatly exaggerated its successes. The Japanese had documented some five hundred sinkings, but only fifty-two Allied boats were known to have been lost. Part of his duty was to set the record straight. He also debriefed the captured American prisoners of war who were now being released from camps in Japan. He recalled that it was impossible to predict who or how many would show on any given day, or where they would come from. For many days, no one came. Then five or six would show up.

The US Navy had to prepare, as quickly as possible, a list of all known survivors from submarines that were lost in action. All personnel had to be accounted for, and the Navy wanted to know how each boat was lost. Paine wrote of this deeply sobering experience and his silent rage over the atrocities committed at the hands of the enemy as told to him by the survivors of the lost boats. Many friends and former shipmates were among the missing, including seven of his thirty-five classmates and several instructors from Annapolis.[19]

Among them was his best friend, Ben Phelps, who had introduced him to the future Barbara Paine when they were all stationed in Perth. He, Phelps, and Bill Mendenhall had become inseparable after they arrived in the Pacific. They went everywhere together. Just before the *Pompon* left on her fifth war patrol, Mendenhall had transferred to the USS *Lagarto* (SS-371). Before embarking, he had asked Paine to hold on to his dress white uniform and keep it clean until he got back. It would stay at the bottom of Paine's duffel bag for the rest of the war. Neither Mendenhall nor Phelps made it off the *Lagarto* when she went down in action in the Gulf of

Siam on May 4, 1945. The fate of his friends' boat was not known during the war. On August 10, 1945, the Navy pronounced it overdue from patrol and presumed lost. Reconstruction of events after the war showed that it was most likely sunk by depth charge from the minelayer *Hatsutaka* in 180 feet of water. Discovery of its wreckage in May 2005 confirmed this.[20]

Another close friend was Bill Hoffman. They had taken their last R and R leave together on Waikiki Beach. Paine had been best man at Hoffman's wedding. Hoffman's boat, the *Herring* (SS-233), had rendezvoused with the *Barb* (SS-220) on May 31, 1944. No one ever heard from them again. Postwar reconstruction of events determined that the *Herring* was likely sunk by Japanese shore batteries off Matua Island on June 1, 1944. In all, one out of every five in Paine's class was killed in action.[21]

He pondered the human consequences of it all, consequences that were all too often deadly for the men and their unseen enemy. The US submarine force sustained the highest mortality rate among all branches of the military during the war. Fifty-two vessels were lost at sea; one of every five submariners was killed in action. Paine wrote in his journal of the unmistakable highs and lows of being a wartime submariner: "The days were dull," as war patrols became all the same, with long periods of monotonous waiting punctuated by the sudden danger of violent and deadly action. "To me it was all fascinating—getting to know the boat, the people, the routine. The romance of piracy under Oriental seas combined with the technological complexity of submarine operations fascinated me. I never got over it."[22] But the disillusionment of war appalled him; it shaped him profoundly.[23]

3

A LONG VOYAGE HOME

Ah! The good old time, the good old time.
Youth and the sea. Glamour and the sea!
—A quote in Tom Paine's wartime journal from Joseph Conrad's Youth

Standing on the pier of Apra Harbor, he saluted his captain goodbye. The scenic natural port on the west coast of Guam provided a picture-postcard setting for Tom Paine's last glimpse of the USS *Pompon*. She was heading home to Pearl Harbor. Her job in the Pacific was done; the war was over. He waved one last time to the crew standing on the bridge. Steering clear of the wreck of the cargo vessel *Tokai Maru* lying on the bottom, the submarine cleared the harbor and headed out to sea. He fixed his gaze on the conning tower until it slowly disappeared below the distant horizon. He turned around, walked back up the pier, and went back to work.[1]

For Paine, the tedious work of cleaning up after the war continued. Through the fall of 1945, the pace of US occupation of Japan intensified. The defeated enemy still had a very viable military, and the US was overseeing its stand-down. The IJN fleet of submarines now needed to be completely demilitarized, and he was ordered to remain behind and make sure that it was done.

He left Guam a few days later on the submarine tender USS *Euryale* (AS-22) and steamed for the Japanese mainland. Now, only days after the surrender, they entered Sasebo Harbor on high alert with weapons ready. He carefully surveyed the port. Through his binoculars, he saw the extent of its destruction, wrought by the thousands of bombs dropped from B-29

Superfortress bombers. The sight and smell of the wretched, burned city and its oily, messy harbor littered with the twisted wreckages of many ship skeletons underscored, for him, the tragedy of the war. Writing in his journal he wondered what insanity had led the leaders of this hidden and now shattered empire to believe that they could take on the mighty United States of America.[2]

Going ashore with the first tender of marines, no one quite knew what to expect. The Japanese had just surrendered, and although hostilities had officially ceased, wartime anxiety was still clearly in the air. He and the other officers all carried sidearms, escorted everywhere they went by the Shore Patrol. Kamikaze resistance, diehard fanatics, snipers—anything was possible. No trouble materialized, however, as the days went on. Only a brief encounter with a small, rogue patrol boat one day interrupted their work. It was quickly squelched by the Japanese officers themselves.

Paine was in Japan to seize a sample of every torpedo that his team could find. It was a priority order that came directly from the Pentagon. During the war, the US had learned the hard way to respect the Japanese torpedoes. While many of the early models had used old German designs, later indigenous variations developed by the IJN were by far the most advanced torpedoes of the war. The search and seizure turned out to be quite successful. Several torpedoes that his team collected are still on display as historical artifacts at the Navy submarine school in New London.

The *Euryale* then sailed from Sasebo and around Kyushu through the Inland Sea to the port of Kure. They sailed along the picturesque, pine-clad coastline that looked like a watercolor painting from a Hiroshige print. It was an eerie sight now, marred by the burned hulks of ugly wrecked ships. Midget-subs abandoned under construction were scattered about the dry docks. Hundreds of bomb craters from American B-29 air raids littered the shoreline. The once impenetrable naval stronghold was now a dilapidated mess of concrete and iron rubble. He was now in the heart of the imperial war machine whence such horrible death and violence came.

His orders were clear and precise: locate and disarm the remaining Japanese fleet, interrogate the crews, search their records, study the materials, and, when the time came, scuttle the boats. To do this effectively, he had to register and disarm each of the submarines that were slowly making their way back in from the Pacific.

One by one they came. Some looked like they had seen little action. Others were so heavily damaged that, as an engineer, he marveled that they were able to make it home at all. Most were of the Kaiten ("Reverse the Destiny") type that carried suicide torpedoes. (The Japanese had turned to these in the last months of the conflict in a desperate attempt to turn the tide of the war.) He watched with compassion as the malnourished crew of each boat climbed ashore one at a time, disgraced and humiliated, and surrendered.

Days turned into weeks. His paperwork was almost complete when a submarine designated as the I-58 appeared on the horizon. As the duty officer on watch, he gave permission for the boat to enter. As it moored at the dock, it looked like most of the others that he had seen, distressed but not heavily damaged. He assembled a boarding party and boarded the sub. They walked by the crew, who wore blank stares as they lined up in their oily uniforms. When they reached the bridge, the commanding officer, whose uniform was cleaner, introduced himself in highly accented but understandable English. He told Paine his name: Commander Mochitsura Hashimoto. Saluting sharply once, Hashimoto then turned around and led Paine's party below to the wardroom. Once there, he and his senior officers bowed low and laid their swords on a table that looked to have been hastily covered with a white tablecloth.

Other captains had offered their commands to him in surrender before, but not like this. Paine refused, by protocol, Hashimoto's offer to surrender his sword. He explained, using the bit of broken Japanese that he had learned while on Guam, that he was onboard only to issue disarmament instructions.

He asked Hashimoto to tell him about the I-58's operational career. The proud Japanese commander stared at him and immediately became irritated, Paine later recalled. After a long, very awkward silence, he looked puzzled, mumbled something in Japanese, and declared that *he had been expecting them*. This was the submarine, he said with no more pauses, that sank the American warship that carried the atomic bomb.

Paine was dumbstruck. He exchanged a look of bewilderment with his lieutenant and asked Hashimoto to repeat his claim. Atomic weapon information was classified. All, by then, knew of the bombing of Hiroshima and Nagasaki that had ended the war. But details such as which ships had transported the components for the bombs had not been revealed to anyone, much less the enemy.

As he began to question Hashimoto, the captain's mood changed. He eagerly unveiled a chart and was soon describing, most animatedly and in great detail, the events of that night. Using a mix of English and Japanese, he told Paine how he had sighted, approached, and attacked the heavy cruiser USS *Indianapolis* (CA-35) in the early morning of July 30, 1945. He had readied the Kaiten suicide torpedoes. But with a clear moonlit sky, a calm sea, and an advantageous position forward of his target, Hashimoto went with conventional torpedoes. Two of the six found their mark. The flagship of the US Fifth Fleet sank in less than twelve minutes, taking with her three hundred souls just past midnight in the greatest single loss of life at sea in the history of the US Navy. Nine hundred survivors battled the ocean, exposure, and sharks for four days, as the Navy failed to notice that she was overdue. The sinking and horrific loss of life that followed was not a highly guarded secret by the time of Hashimoto's encounter with Paine. But at the time, few knew that the *Indianapolis* was the ship that had delivered the radioactive cores for the Hiroshima and Nagasaki bombs to the island of Tinian or that the I-58 was the one that had brought it to the bottom. This was the first time that anyone had heard Hashimoto's fantastic story in person. Still somewhat stunned, Paine escorted the captain out onto the deck of the boat and briskly walked him over to the base command post, where Hashimoto expected to be questioned some more. Paine never saw him again. After the war, the US would summon Hashimoto to Washington to testify against Captain Charles McVay of the *Indianapolis* at his court-martial. Hashimoto passed away in 2000. In 2001, McVay was posthumously cleared of all wrongdoing by the Navy.

Days after Hashimoto surrendered his boat, Paine drove to Kure and saw for himself the destruction that the atomic bomb, a uranium fission device called *Little Boy*, had wrought. In the wink of an eye, it had obliterated the once picturesque, 350-year-old city of Hiroshima and the smaller town of Kure. The ruins stretched on for miles. A ghostly, colorless ash cloaked everything he saw. Even the rain had a strange odor. Occasionally, a sick person would walk by. Another would wander pitifully among the rubble. Some peddled by on bicycles, their shoulders burdened with heavy water jugs. The somber sight was etched deeply in his memory.[3]

In early November, his new commanding officer, Commander James E. Stevens, had him group all the captured Japanese submarines together at Sasebo. He gave orders for them to be ready to get underway on four hours' notice. The war had now been over for two months, and most of the Japanese submarines had been either demilitarized or scuttled. Many were, however, still moored in the harbor. The US had yet to decide on their disposition. As the acting division commander of Submarine Division 2, Paine commanded seven of the captured subs.

With his Japanese now slowly improving, he was able to work as smoothly as could be expected with his chief lieutenants, Murayama and Takezaki. Both spoke a bit of English. As the days wore on, a common but wholly unexpected bond formed between him and the Japanese officers. He was learning quite a bit about his former foes. Some information was militarily very important, like the tactics and patrol chronicles that each boat went through. Others were of more human interest. He learned first-hand from Murayama of the overwhelming, tearful emotions that the crew and officers shared as they launched the Kaiten suicide torpedoes. Other stories surprised him. For example, Takezaki admitted that his side had greatly overestimated how well their midget submarines might work in the first hours of the attack on Pearl Harbor. They should have been far more effective than they actually were. He also revealed to Paine the surprising fact that, as far as he knew, Japan never suspected that the US was successfully decoding their messages during the war under the covert Office of Strategic Services program code-named Operation Magic.

As they conversed, both he and the Japanese began using terms like "us" versus "them" to describe submarines versus surface ships—not Japanese versus American. His surprise at how quickly the bond formed from their shared experience so soon after the end of hostilities remained with him long after the war.

Most of the boats slated for scuttling that winter off Goto Shima, just off the western coast of Kyushu, had been gathered and moored in Sasebo Harbor by mid-November. They were mostly of the small, tactical variety that the Allies had seen plenty of during the war. But drawing very high interest were the gigantic boats of the I-400 Sen Toku ("Special Submarines") class. The 5,500-ton leviathans were 400 feet long and nearly

40 feet high. Each had a crew of 145. Each aircraft-carrier submarine could deploy three M6A1 Seiran ("Mist from a Clear Sky") specially designed aircraft. These could be folded up like origami and transported inside the thick-walled hangar of the submarine.[4]

Vice Admiral Charles A. Lockwood, commander of US Pacific submarine operations, had revealed to the fleet squadrons after V-J Day the never-completed mission of these massive super-subs. Parked off the US West Coast, their mission was to infest the North American continent with rats and mosquitoes infected with bubonic plague, cholera, and other agents of biological origin. During the war, chemical and biological weapons of mass destruction had been secretly developed by General Ishii's infamous medical experimentation laboratory in Harbin, Manchuria. There, the Japanese committed war crimes on a wide scale by injecting Chinese prisoners and civilians with awful contagions as part of its laboratory for human experimentation.

In March 1945, the Japanese Army General Staff changed the mission of the I-400 to one of bombing the Gatun Locks at the Panama Canal. This too was never carried through. In June, the super-subs were ordered to the Ulithi Atoll, where the American task force was assembling for the planned invasion of Japan. They then finally turned around and returned to the homeland in August after Japan's surrender.

With an unheard-of 37,500-nautical-mile intercontinental cruise range, the Sen Toku were more than just another object of curiosity in the Axis arsenal of secret weapons. They were the largest submarines in the world. Their size and speed would be unmatched until modern-day nuclear submarines came onto the scene in the 1960s. With the advent of the atomic bomb, the US naval high command thought that the submarines could be used to launch ballistic missiles at sea. They wanted to see the submarines for themselves. It was up to Stevens and his men to bring them back to the US.

Paine was ordered to report to Commander J. M. McDowell as the executive officer on the I-400.[5] Sailing a captured, classified submarine across the Pacific required a good amount of preparation. With no blueprints or instructions to work from, he had to be creative. He ordered the crew to salvage whatever parts and supplies they could scrounge from the remaining warehouses and caves around the shipyard. Since there was no plan to submerge the boat on the voyage back to the US, a lot of work was saved by not having to repair the snorkel and diving system. He made sure the sub was stocked with enough provisions for fourteen days. They had to first get to Guam. The boat would then be resupplied for the next leg of the voyage to the Marshall Islands, before the long voyage to Pearl Harbor. The men filled the cavernous hull with all sorts of war trophies: rifles, bayonets, Japanese uniforms, and even a seaworthy sampan. He kept a shiny, antique samurai sword for himself.

The trio of submarines (the I-14, I-400, and I-401) got underway for Guam on December 11, 1945. Paine managed to get out a quick telegram home before embarking: "CAPT G T PAINE USN 450 OCEAN AVE SEALBEACH CALIF MERRY CHRISTMAS EASTBOUND JAP SUB HOME SOON TOM USS EURYALE"[6] They headed southeast out of Japan, slowly at first, to steer well clear of the unswept minefields west of Kyushu. They passed through the Tokara Gunto, a passage south of Kyushu that was often the scene of heavy combat. Upon reaching the open waters of the Pacific, it was full speed ahead.

Up in the conning tower, he no longer scoured the sea for enemy mastheads and periscopes. When he went to the bridge at twilight to get a star sighting, he stifled the urge to douse the running light. The pleasure of peacetime sailing began to sink in. He wrote in his journal in large letters: "WE OWNED THE SEA!"

Nearing Guam, McDowell was nervous about the keel clearance of the massive boat. A new pipeline had just been laid in the shallow channel of Agana Harbor that was not marked on the charts. But earlier in the war Paine had received his Second Class Deep Sea Diver certification there and knew the variances of the muddy terrain. McDowell handed the helm over to him. Twenty minutes later, the I-400 was in port.

After a brief layover for R and R, the squadron continued on. They took an easterly course to the Marshall Islands, two thousand miles away. Christmas Eve saw them carefully navigating the low-lying reefs around Eniwetok Atoll, the site of many future nuclear weapon tests. McDowell did what he could to keep morale up on the long journey home. He posted a poem in the control room that evening after dinner. Paine jotted down the words on a scrap piece of paper that he kept for the next forty-five years:

"Christmas at Sea, 1945"

A Merry Christmas, which I know
is better here than in Sasebo!
Next Christmas, and the ones to come,
I hope all hands will spend at home.
Let's hope and pray that ne'er again
must we spend Christmas killing men.
That peace will reign beyond our time,
no guns compete with Christmas chimes.
Let's offer thanks for where we are,
for Christmas time not spent at war,
and honor those who gave their lives,
while we head home toward our wives.

Commander J. M. McDowell, U.S. Navy
Commanding Officer USS Ex-HIJMS I-400

After refueling at Eniwetok, they embarked on the last leg of the journey to Hawaii. On January 6, 1946, they arrived at Oahu and entered Pearl Harbor. As they glided quietly by the sunken hull of the USS *Arizona*, the crew solemnly dipped the US and Japanese ensigns.

There was immediate curiosity about the colossal aircraft-carrying submarines. But the interest waned considerably after some initial investigation. The twin-hull design was unconventional and impressive, but the Americans did not find the Sen Toku nearly as advanced as they had thought when they first received word of them. Paine, however, was not ready to give up. He thought that the subs were a marvelous feat of maritime engineering. He prepared a highly detailed, classified briefing for his commanders and briefed anyone who would listen. He urged the Navy to refit the submarine so its diving and submerged performance could be evaluated. Paine relished the thought of taking her down and seeing how well she would perform at operational depth.

But after finishing their initial assessment, naval intelligence was no longer interested. After they got all the information they needed, the

I-400 was towed out to sea and scuttled unceremoniously with four torpedoes. (The other two boats met the same fate.)[7] The Cold War was now on the horizon, and the Navy knew that the Soviet Union would be just as curious about the Japanese secret weapon as the US had been. As far as the Americans were concerned, the wartime relics were best left at the bottom of the ocean.

On their long voyage home, Tom Paine had told McDowell that it would be his last submarine command. He had decided not to reenlist. He could have stayed in the Navy, however, and in fact, had a few different options available. After his fifth war patrol on the *Pompon*, he had applied for postgraduate studies in naval architecture at the Naval Postgraduate School in Monterey; he was selected as an alternate for the assignment. At the time, he wrote in his service résumé that, "I am interested in the regular Navy as a career."[8] On his discharge papers, McDowell recommended that Paine be promoted to the rank of Lieutenant Commander. While he would have been in line for his own command, he told McDowell that peacetime training would probably be a big letdown. McDowell understood; he did not disagree.[9]

Writing to his parents from Oahu while recovering from a reaction to his second cholera shot, Paine said that he was ready to come home. He wanted to go back to school and learn about all the new technologies that had come out of the war. "When I get back to the [S]tates, I'm going to decide, in a general way, the things I want to start at after the war. . . . It will probably involve more school, perhaps at Cal Tech? We'll see."[10]

In February, he sailed on the USS *Boarfish* (SS-327) from Pearl Harbor to San Diego. From there, he went home to Seal Beach. On March 24, 1946, Tom Paine was honorably discharged from active military service. During his two years in the Pacific, he had earned the Submarine Combat Insignia with Two Stars and a Pacific Fleet Commendation Ribbon with Combat Clasp for Performance in Action. For the next fifteen years, he remained on inactive duty and served in the Naval Submarine Reserve on the Scientific Reserve Units of San Francisco, Schenectady, and Boston. On December 1, 1961, Lieutenant Paine ended his seven years, two months, and twenty-two days of service in the United States Navy.[11]

Trained as a secretary, Barbara Helen Taunton Pearse had joined the Royal Australian Air Force when she was just eighteen. The future Mrs. Paine grew up in a working-class Australian family. Her mother, Marguerite Jones Pearse, had died at the age of forty. Her father, Henry William Taunton Pearse, sent what little money he could from working days and nights in horrible conditions as a lighthouse keeper on barren Rottnest Island, a longtime penal colony off the coast of Western Australia.

She had spent most of the war supporting the New Guinea campaign as a plane spotter and ground controller.[12] Barbara and Tom had met in Perth in 1943 when Paine was first stationed in Fremantle. But he had been unable to go back to Australia after the *Pompon* left the base on her fourth war patrol in February 1944. When he reached Guam after the surrender, he asked naval command there to help him find a way back to Perth so they could marry. But McDowell told him that all US nonessential operations in Australia had been halted. He wrote Barbara the disappointing news: she would have to find her own way to America. He wrote home to his parents that "Barbara and I are determined to wait forever if need be."[13]

Following the war, there was an enormous waiting list of Australian war brides who were left to anxiously apply and make their way to the United States. After being discharged from the RAAF on October 2, 1945, she waited a year and a half before her name was finally called. With the money she had saved from the war, Barbara Pearse caught a Matson Line steamer to America in the fall of 1946. With the polished samurai sword he had buccaneered on the I-400, Tom and Barbara Paine cut their wedding cake on a sunny southern California Tuesday afternoon, October 1, 1946, in a traditional naval ceremony at the Long Beach Naval Shipyard.[14]

4

HOUSE OF MAGIC

Any sufficiently advanced technology is indistinguishable from magic.
—*Arthur C. Clarke*

The panorama along the winding curves of the Pacific Coast Highway stretched on for miles. Going from Seal Beach north to Palo Alto finally gave him a chance to relax, gather his thoughts, and move on. The war that he had fought on the other side of the ocean was behind him. Like others who had endured and lived, he was still readjusting to life back in the States. The young were coming home to a changed postwar America filled with more hope, opportunity, and optimism than ever before. Tom Paine was ready for the challenge.

Returning from the Pacific, he applied to several graduate schools around the country: Brown, Columbia, MIT, and in-state at nearby Cal Tech and the University of California at Berkeley. He wanted to really understand engineering, but thought also of architecture. Although his undergraduate grades at Brown had only been fair, his wartime service record and recommendations had been sterling. It was enough to gain him admission to Stanford University. In the fall of 1946, he enrolled in the School of Engineering and entered the mining and metallurgy program. In October, Barbara joined him. They moved into a one-room apartment in an old Palo Alto hospital near the campus. With an influx of postwar students enrolling on the GI bill, the school had converted it to a dormitory for childless, veteran couples.

Highly technical courses such as Mining, Geology, Industrial Process, and Pyrometry filled his days. He liked these subjects, and received mostly As and Bs. Others, like Russian, were not quite as appealing. Although it was a difficult language to learn (he received a C), he recalled thinking that knowing some Russian might turn out to be useful in the world of international commerce after the war.[1]

He immediately began work on the Navy's early nuclear reactor program as part of his research. In the years following World War II, the US Navy was in a tight race with Great Britain and the Soviet Union to develop the first wholly contained nuclear reactor. A reactor sealed inside a pressure vessel could theoretically power a ship or submarine for decades. Such a vessel would be able to stay at sea for years without refueling. The security clearance he had from the War Department and his wartime experience on the *Pompon* made him an ideal choice for the Navy research program.

Under O. Cutler Shepard, a pioneering metallurgist and the department's distinguished professor, Paine conducted laboratory research whenever he was not in class. His thesis research was to find out how liquid metals could be made to interact in a useful way with high-temperature alloys. A nuclear reactor requires a great amount of cooling. The Office of Naval Research had asked Shepard to see if liquid metal could be used (instead of water, for example) to cool a reactor. Led by Admiral Hyman G. Rickover, the reactor program was one of a number of highly classified postwar Navy efforts involving the universities, and one of the largest. Eight students under Shepard would receive their PhD degrees by working on the project.

Paine recalled finding the highly theoretical work "intensely interesting." The hours were long. He often sat alone at night performing experiments. The university had a stress-rupture machine in its Mineral Sciences Laboratory that was his primary research tool. Paine became quite proficient at using it to test the compression, shear, and tensile limits of different kinds of materials. Using the powerful lens of an electron microscope, he scanned for traces of cracks and fractures in a variety of steels and alloys.

Stanford did not pay him much, just $110 a month. An additional graduate student subsistence allowance of $90 made a big difference. It was enough for him and Barbara to get by. After two semesters, he received his Master of Science degree in Physical Metallurgy on June 15, 1947.[2]

He continued to study for his doctorate. In his second year, Shepard made him his lead graduate student. For the next two years, he wrote papers on the results of his research; several written with Shepard were published in classified Department of Defense literature.[3] In April 1949, he passed his comprehensive examination in General and Metallurgical Engineering. This put him into the School of Mineral Sciences PhD program. The Stanford University Committee on Graduate Studies approved his dissertation on May 20, 1949. Ten days later, he successfully defended his thesis, a technical treatise titled "The Effect of a Molten Lead Bismuth Eutectic Alloy on Steel."[4]

He now had a decision to make. A doctorate in the very specialized field of Physical Metallurgy meant he could either teach at a university or work in the burgeoning high-tech industry. Shepard asked him to stay, and offered him an accelerated professorship in the department. Paine wanted to teach, but just not yet. He recalled telling Shepard that he thought it was important, before teaching someone, to "first acquire a broader level of practical work experience for himself." So he reluctantly turned his professor down.[5]

In its heyday, General Electric was the preeminent technology company in the world. No other had quite the cachet of GE. Young engineers and scientists came from all parts of the industrialized world to try and make their mark at the flagship company of modern high technology. Paine was no different. In the fall of that year, he and Barbara packed up and headed east to Schenectady, New York. That was where Thomas Edison had first set up shop in 1892. On October 1, 1949, he joined the GE Research and Development Center as a research associate. Throughout the industry, the campus was known as Schenectady Works, or "The Knolls." To the public, it was a wondrous place of modern marvels at the dawn of the space age. Many simply called it "The House of Magic." GE had it all: electric blenders, washing machines, fluorescent lights, refrigerators. Every Sunday night for ten years, millions gathered in their family rooms in homes all across America and watched the *General Electric Theater* hosted by a Hollywood actor named Ronald Reagan on live television. Everything that made for the perfect image of American suburbia of the 1950s was embodied in General Electric.

Paine was now a new but highly trained materials engineer. For the next twelve months, he experimented on the magnetic properties of unusual metals. The work was very specialized. Hours of laboratory work were needed to return one data point. His team discovered that strong, man-made magnets could be formed by mixing iron-cobalt with lead powder. The magnets had a lot of uses. GE engineers could mold them into any shape they wanted using a process called powder metallurgy. The magnets could then be used in everyday microscopic applications. These ranged from hearing aids to automotive test equipment. His work led to GE's patent on the Lodex permanent magnet. It has since been continually used in the commercial automotive, electronics, and communications industries.

When Paine first started, J. Herbert Hollomon, his supervisor, had given him a verbal promise. Hollomon was the laboratory's assistant manager for metallurgy research. He told Paine that after being in New York for a year, he would have the chance to set up his own section at the Meter and Instrument Laboratory in the company's Lynn, Massachusetts, facility. There, he could run his own programs and branch out into new areas of research. Hollomon kept his word. On December 1, 1950, he signed Paine's transfer papers and sent him 150 miles east to Massachusetts.

Paine continued the research on fine-particle magnets while in Lynn. The work led to production of GE's first one hundred thousand Lodex magnets. He and his chief scientists, L. I. Mendelsohn and F. E. Luborsky, ran and grew the laboratory. Over the next five years, the quality of its research reached new heights. In 1956, it received the prestigious American Association for the Advancement of Science award for Outstanding Industrial Application of Science. By 1960, the Lodex magnet was well on its way to becoming an industry standard, as annual sales surpassed $1 million.[6]

When his supervisor, M. A. Princi, left on November 1, 1955, to take a job as the GE executive engineer in Milan, Italy, the company promoted him to general manager of the Lynn Instrumentation Laboratory. At thirty-four, he got his first taste of directing a large technical organization.

The focus of his laboratory was to support the Lynn facility in its primary job of making jet engines. In the 1950s, jets revolutionized transportation and defense at an astonishing pace. Between them, General Electric and Pratt & Whitney made virtually all of the jet engines in the United

States. His laboratory found ways to use leading-edge metals to improve engine performance at higher temperatures and pressures. The first of some two dozen patents he would receive was for a time-temperature integrator control system. GE used it to test aircraft ranging from the popular civilian Learjet to the super-sleek B-58 Hustler strategic supersonic bomber.

He also continued his work on shipboard nuclear reactors. The laboratory developed a first-of-its-kind solid-state instrument that could monitor the temperature inside a nuclear reactor. The US Navy would use the invention in its fleet of nuclear submarines and ships throughout the 1950s and 1960s. The USS *Enterprise* (CVN-65) was the most recognizable of these ships. It was the world's first nuclear-powered super-carrier and was just then becoming operational.[7]

Back at Stanford, Shepard still wanted him back. But Paine was quite content now. Times were good for Tom and Barbara. They now had four small children: Marguerite, whom they called Greta, was seven; George, four; Judy, two; and their youngest, Frank, had just been born. They thought they just might stay a while in the Massachusetts Bay community. He wrote to Shepard, saying, "General Electric has been very good to me and can promise an interesting future."[8]

It was late in the afternoon. Up until he heard the news, October 4, 1957, had been much like any other Friday. He was at his desk getting ready to go home when someone told him to come over to the lounge and listen to the breaking news coming over the radio. The announcement soon silenced everyone in the room. The announcer said that the US Air Force had just confirmed that the Soviet Union had launched the world's first artificial satellite into orbit. They were told that in the early morning sky the next day, they would be able to see the light from the satellite that the Soviet Union called *Sputnik*.

Before sunrise the next day, he and Barbara put jackets on the children and went down to the beach. Standing on the shore of the Atlantic, they gazed into the sky and waited. Then four-year-old George suddenly looked up and yelled, "There it is!" They looked to where the boy was pointing and saw *Sputnik* gliding effortlessly across the dawn's morning sky. He remembered feeling only a stark sense of awe.[9]

Ralph J. Cordiner was an influential businessman. In 1958, the iconic CEO of General Electric ordered the complete decentralization of the vast high-tech company. He separated the various groups of the corporation and put product line responsibilities under a brand-new management structure. The change streamlined GE and saved millions of dollars while boosting its stock value. But part of the savings came by way of eliminating management layers and cutting the number of mid-level managers. Paine was one of them; he found himself with no choice but to leave Lynn. In September of that year, he transferred back to Schenectady. Herb Hollomon was waiting for him, this time in the Metallurgy and Ceramics Research Department.

Studying the behavior of composite materials in a time before there were digital computers was not easy. Solutions to "finite element" analyses of materials that are solved entirely by computers today took hundreds of hours to do by hand. Paine performed thousands of calculations using a slide rule and plotted the data on translucent onionskin graph paper with a wooden pencil, all by hand.[10] The tedious work supported other parts of GE. The applied research he performed had one goal, and that was to improve the commercial products the company made and sold to consumers around the globe.

In the spring of 1960, Hollomon left for Washington, DC, to become assistant secretary of commerce for science and technology in the new Kennedy administration. This left Paine in charge of the materials lab. Paine liked Hollomon, but was glad he left. They had worked closely together for many years. He considered him a fair mentor, a good technical manager, and an excellent engineer. He recalled learning most from Hollomon about the nuances of GE's complicated partnership with the federal government. But he did not always endorse Hollomon's inflexible way of doing things. Now, he was in charge, and had his chance to run the lab as he saw fit.[11]

He thought the laboratory had underperformed with Hollomon as manager. To reach its full potential, it needed to broaden its customer base and win new government prime contracts by relying on the laboratory's proven technical merit and past performance. Paine received permission from the corporate office to branch out to reach a more diverse set of clientele from various agencies of the federal government. He brought in new, nontraditional GE customers such as the National Bureau of Standards and the US Geological Survey, agencies that would go on to use the lab's expertise in unorthodox and esoteric ways. They looked to his lab for quick results. Projects became much more dynamic and were no longer limited to research and development in material science. Work in medicinal electronics, water purification, and urban transportation now complemented the other engineering programs. GE headquarters began to take note. At age forty, Tom Paine was starting to make a name for himself.

In August 1956, General Electric had opened an office in the quiet seaside community of Santa Barbara, California. Nestled in the narrow range between the steep Santa Ynez Mountains and the Pacific coast, it was a new kind of office, gathering in one place many of the company's top engineers and scientists from around the country. Their job was to use science and math to forecast the fast pace of change in the world of high technology. It was officially called the GE Center for Advanced Studies, but most people just called it TEMPO.[12] Over the next couple of years, the office grew and became the center of excellence for GE's national defense research business.

East-West tension between the US and the Soviet Union was at an all-time high in the late 1950s and early 1960s. One miscalculation on either side could have led to an all-out nuclear war. Civil defense exercises were part of daily life for school children across America. Backyard bomb shelters were not uncommon. It was in this precarious Cold War setting that TEMPO had the very difficult job of trying to predict the needs of the country fifteen years into the future. They had to come up with creative solutions to better the country's military, recommend ways to grow the national economy, and actually predict the future without being too hyperbolic. Experts gathered intelligence from government and private sources around the world and from classified materials in order to brainstorm the trends and possibilities in technology. What was the trajectory of the arms race, what was just over the horizon, and what were the changing needs of national security? Key in all this was selecting weapons (primarily nuclear) that would guarantee America's survival in a protracted Cold War.[13]

Richard C. (Dick) Raymond was TEMPO's first general manager. He had set up the center as an intellectual community that the Department of Defense could call on at any time for special studies and advice. He realized early on that to operate effectively as a "think tank," they had to have as much autonomy from the rest of GE as possible, both geographically and in terms of the makeup of its people.

Raymond staffed the office with experts from all walks of life. Linguists, psychologists, and economists sat next to mathematicians, scientists, and engineers. Unlike other parts of General Electric, TEMPO delivered no products, only ideas. In just a few years, the Santa Barbara operation became one of the top think tanks in the country. With national goals and priorities changing all the time, the phone was constantly ringing. Washington wanted ever-more visionary solutions to difficult problems, and it wanted them fast.

But by 1960, the group was beginning to struggle as a profitable business unit. In 1961, TEMPO had a major role in the disastrous cancellation of the B-70 Valkyrie strategic bomber program. At the time, the program was one of the largest Defense Department acquisitions since World War II. Designed by North American Aviation in nearby Downey, California, the Valkyrie was a very large, Mach 3, six-engine bomber that could fly well over 70,000 feet. This would have made it invulnerable to the MiG-21 interceptor—at the time the only Soviet defensive capability against the bomber.

TEMPO studies had pointed out, however, that Soviet surface-to-air missiles had advanced to a point that high-altitude bombers were vulnerable. A particular threat was the S-75 Dvina (code-named the SA-2 *Guideline*) missile that could fly in excess of 80,000 feet to bring down an aircraft. This effected a fundamental defense policy change with regard to ballistic missiles and strategic bombers. The Atlas intercontinental ballistic missile (ICBM), which was already in production by the Convair Division of General Dynamics, could also deliver nuclear weapons anywhere into the vast Soviet territory. That March, President Kennedy canceled the expensive bomber. The irony was that GE was already under contract to build the aircraft's engines. Executives at headquarters in New York were furious. They demanded a change in Santa Barbara.

Paine was looking for a change, too. He recalled that he had reached an impasse with the senior leadership in Schenectady. "It's hard to get emotional about the GE monogram, and it's not always a place where you can find the crisp intellectual life. The question is whether you want to devote your life to devising the perfect watt-hour meter or to 37.2% of the electric toothbrush market. It is a model of a rigid hierarchy, though to be fair, the top men try to do something about it, to achieve a more flexible structure. The trouble is that someone always gets alarmed, and the effect is to turn power back to the pope."[14]

He knew of the situation at TEMPO and applied for the job. He told his managers that "for family reasons, my geographical preference is for northern California."[15] It was not northern California, but it was close enough. In March 1963, GE headquarters appointed him general manager of the Santa Barbara office.

The first thing he did when he arrived was to change the way TEMPO did business. He started with the office building itself. Since it first opened in 1956, employees had been working out of a rented, run-down hotel building in a rather unsavory part of town. He moved the office into the much more attractive Barbara-Balboa-El Presidio historic downtown district.[16] Then he changed the outfit's business strategy. "We concentrated on areas I thought were important," he said. "Rural development abroad, urban rehabilitation here, communications, transportation. What we would say was, 'We have been spending a lot of time looking at the world of the future, and we think we can tell you a lot of valuable things about the problems you will be facing 10 or 15 or 25 years from now.'"[17]

The corporate office still treated TEMPO as somewhat of an outcast. The bearded, open-collar, "West Coast" look of many of the employees did not sit well with the suit-and-tie corporate management back East. Changing that perception was not easy. Trying to restore TEMPO's standing took more effort than he had expected, he told close friend Ed Schmidt. But Gerald L. Phillippe, the President of GE, believed Paine when he said that TEMPO complemented the rest of the company and added value to the corporation. During that first year, he spent a third of his time at the corporate office in New York. The board constantly wanted to know from him how he was going to align TEMPO's business objectives with the rest of GE. Another third of his time was spent on-site in Santa Barbara. The rest of the time he visited customers at different places around the country.

Paine worked with some two hundred experts who were concerned with what they thought would be the condition and needs of the country a decade or so into the future. Many predictions had to do with the condition of the cities—a topic that was seemingly escalating in importance with Washington on a weekly basis. Under President Lyndon B. Johnson, the new Democratic national leadership had a lot of questions that involved urban planning, urban renewal, and modernizing the inner cities. Most of the interest directed at TEMPO, however, still had to do with national security. TEMPO thinkers created "what-if" scenarios and projected their possible impact on the United States. An issue might be how a major breakthrough in the field of disarmament would affect the country. Perhaps a critical trade secret might be compromised. Another scenario was how vulnerable US cities would be to a surprise attack by a devastating secret weapon.

His group forecast that communist China would detonate a nuclear weapon some time between 1963 and 1965, most likely in late 1964; the actual explosion occurred in October of 1964. This brainstorming on the direction of US national security accounted for some 75 percent of TEMPO's work under Paine.[18]

In 1970, *Time Life* correspondent Robert Sherrod asked him what was so special about the group and the work that they did. He sat back, reflected briefly, and recounted a rather unusual story that had Sherrod smiling by the time it was over.

His longtime friend, Ed Schmidt, was an independent consultant who worked for TEMPO. Schmidt was an eccentric, in terms of both personality and profession, a diversely educated, modern-day Renaissance Man. He had degrees from Georgia Tech and MIT, and had an unusually broad range of knowledge on everything from the technicalities of civil engineering to the nuances of the effects of foreign trade variances on domestic affairs. Paine found him very pragmatic and practical, and used him as an adviser, confidant, and sounding board partly to help in his own thinking. They spent many afternoons in Paine's office tossing around ideas. This sometimes resulted in unconventional ways of doing things.

In the early 1960s, the US was trying hard to provide aid to the Republic of Yemen. Yemen was one of the poorest and most unstable countries in the volatile Middle East. But its location at the mouth of the Red Sea made it uniquely important geographically. For this reason, the Soviet Union and communist China were also giving it large-scale assistance.

The State Department asked TEMPO to take a look at what they called "the worst foreign aid situation in any country" in the world. Washington had installed a radio station there only to see the Egyptians capture it to transmit anti-American propaganda year after year. Food shipped by the Red Cross was not reaching the Yemenis. Supplies entering the country had to first dock in Soviet-controlled ports, where they were stolen. They were then transported on a road where they were further pilfered by Chinese communists. The same thing happened at roadblocks set up by the Yemenis. Local warlords helped themselves to what they could. By the time the trucks arrived at the American Embassy, only a sack of grain was left. One sack was always left to encourage the US to try again.

Paine's office could not come up with a good solution for the State Department. Analysts had no answers, and the program dragged on and soon got Paine's personal attention. He took the project over from the program manager and began talking directly with the State Department. One day as they discussed the problem in his office, Schmidt looked out the window and asked, "Tom, why not send me there?" He turned around; Paine looked at him with a big grin. His experts had worked for months to a dead end trying to get intelligence on the situation. They needed someone on the ground. So Paine said why not, and sent his friend into the middle of a war zone. Schmidt talked to the villagers and gathered information from the locals on the supply route, conditions of the roads, and the strength of the resistance. He scouted locations where the US could implement further aid projects. A week later (after claiming that he had dodged a hand grenade, no less), he came back with exactly the kind of information that Paine needed. He wrote up a report and hand-delivered it to Washington. He told Sherrod that it was a rather unorthodox way of getting something done. But it worked.[19]

By the fall of 1967, Paine had been with General Electric for seventeen years. A year earlier, NASA had brought its second manned spaceflight program, Project Gemini, to a successful conclusion with the splashdown of Gemini 12 in the Atlantic. It was the final flight of NASA's two-man spacecraft program. By now, the United States had been launching astronauts into space for six years. Space probes were also venturing near Mars and Venus for the first time, sending back tantalizing black-and-white pictures of Earth's nearest planetary neighbors. Four robotic Surveyor spacecraft had already soft-landed on the surface of the moon. America was now building the most complex rocket and spacecraft ever made in an attempt to land the first human beings on the moon. "Go Fever" was in the air. The country was in a full-blown race with the Soviet Union to put a man on the moon. NASA was in the news and making headlines. While TEMPO had a few small study contracts with the agency, he himself had nothing to do with them.

That was all about to change.

5

IT'S A PRESIDENTIAL APPOINTMENT?

Now, why do you want to do this?
—James Webb to Tom Paine in January 1968

Paine went to NASA by way of a tragedy. On January 27, 1967, a fire at Cape Kennedy's Launch Complex 34 killed the crew of the first mission of the Apollo program. Gus Grissom, Ed White, and Roger Chafee were rehearsing a launch countdown for their scheduled February flight when a flash fire broke out, trapping them inside their capsule. As one of the country's "Original 7" Mercury astronauts, Grissom had been the second American to go into space five and a half years earlier, and White, one of the best-known astronauts in the space program, had been the first American to perform an extravehicular activity (EVA; space walk) in June of 1965. Chafee, who had been the first from the third group of NASA astronauts selected in 1963 to be assigned to an Apollo mission, was about to make his first flight into space. Their deaths shook NASA and stunned the nation.

The accident completely devastated James E. Webb, the Administrator of NASA. Webb was a powerful administrator. The ex-Marine aviator turned lawyer was experienced, reliable, and quite politically savvy. He had already been in the Washington establishment for nearly thirty years when he took over the reins of the space agency in 1961. Having been the director of the Bureau of the Budget and later the undersecretary of state under President Harry Truman, he was one of the longest-tenured administrators in the entire federal government. Perhaps most importantly, he was the person who had sold the fledgling space program to John F. Kennedy in the early days of his presidency.

In the aftermath of the fire, Webb had put the blame squarely on North American Aviation, the prime contractor of the Apollo command and service modules.[1] The spacecraft that was designed to fly astronauts to the moon and back had instead killed three astronauts before it ever left the ground.

In his mind, people inside NASA had also failed him. Findings of the space agency's internal accident inquiry board and subsequent congressional hearings strongly supported that conclusion. The findings also supported Webb, but only to an extent. They underscored, in particular, an alarming pattern of steadily deteriorating leadership that had been going on unbridled for some time within the agency. Contractors testified that orders from Headquarters in Washington contradicted other orders coming out of field centers in Texas, Alabama, and Florida. Loose leadership in the upper ranks of the Apollo program had gone unchecked. Worst of all, safety had been compromised.[2]

Webb believed all along that the Apollo 1 fire would not have happened had his people been more observant and fastidious. Even before the hearings were over, the purging began. Over the next ten months, he implemented a sweeping reorganization at NASA Headquarters, such that by the beginning of 1968, he had replaced nearly every high-level administrator reporting to him. The most important of these moves had to do with Robert C. Seamans Jr., his deputy.[3]

Informally dubbed "The Front Office," the Office of the Administrator is the top management tier of the space agency. It consists of the administrator, deputy administrator, associate administrator, and staff that support them. All of NASA ultimately reports to this office. This includes the various associate and assistant administrators at headquarters and the directors of the field centers from around the country.

Not quite two years earlier, in December of 1965, Hugh L. Dryden, the longtime deputy administrator of NASA, had passed away after a brilliant, five-decade career. Following Dryden's death, Seamans was promoted from associate administrator to inherit the position of deputy administrator at the agency. He became Webb's right-hand man, but he did so only reluctantly and at Webb's insistence.

As Webb's deputy, Hugh Dryden had been a tour de force. It was said of him that if something needed to be done in the scientific establishment of the United States and he agreed that it should be done, Dryden could make three telephone calls and it would happen. But his greatest contribution to the space program may have been to bridge the seemingly unbridgeable chasm that had developed between Webb and Seamans, the third-ranking person in the agency.[4]

In his autobiography, *Aiming at Targets*, Seamans described his relationship with Webb during this time as slowly taking a turn for the worse. It became irreparable following the Apollo 1 fire, deteriorating in a most precipitous way.[5] Seamans eventually saw no point in staying at NASA and decided to move on. He tendered his resignation on October 2, 1967. In a somewhat unexpected move, however, Webb asked him to stay on as a consultant until he could find a successor. Webb knew that Seamans had a well-established reputation as the agency's most eloquent spokesman in Washington on technical matters and was held in very high esteem by the space community. Despite their falling-out, Webb knew that he could not afford to have Seamans simply walk away. So monumental was Project Apollo that if it failed, no amount of political posturing was going to save NASA.[6]

The space agency now needed a new deputy administrator. This was how Paine entered the picture. The person who linked him to Washington was John W. Macy Jr., chairman of what was the Civil Service Commission (CSC) under Presidents Kennedy and Johnson,[7] later reorganized into the Office of Personnel Management. Since the deputy administrator of NASA is appointed by the president, the White House had tasked Macy to find someone to replace Seamans soon after his resignation. As the chairman of the CSC, Macy was a powerful man in Washington and extremely well connected. He knew every high-level appointee in Washington, and had a network of people at the federal, state, and local levels who maintained files on all persons considered for top jobs in the federal government.

It turned out that Macy already had a file on Paine. This was not mere coincidence. The year before, the CSC had approached Paine for the position of assistant secretary of transportation for research and development. He had declined, however, after Macy's office put him in touch with Alan S. Boyd, the secretary of transportation. After speaking with Boyd, he had decided not to pursue the opportunity. At the time, the job appeared to him to have been structured in such a political way that it would not have allowed him the autonomy to get things done within the hierarchy of the Department of Transportation.

It was different this time around. Paine fit the bill. He was the nonpolitical, noncontroversial type of technical manager that NASA wanted to bring in from the industry following the fire. The federal government as a whole was then actively seeking to recruit more experienced people from the private sector. Macy remembered Paine from his connections with Boyd and Herb Hollomon. He looked at Paine's file again, this time with the NASA job in mind. Macy knew the particulars of this appointment quite well. One of the duties of the deputy administrator of NASA was to handle all high-level personnel negotiations with the CSC for the space agency. Paine's record looked good, as did his political party registration (Democrat).

Paine would later remember that when Macy's office called him in Santa Barbara, he was not rapt. The title "deputy" sounded rather unimpressive, suggesting not much more than another government billet that had to be filled.[8] He did not know much about NASA other than what was in the news. In all his years at General Electric, in fact, he had not worked on any NASA contract of consequence. The conversation was not long, since Macy's secretary did not have a lot of information for him. However, since he was going to be in Washington on business the next day, he told her that he would stop by and talk with Macy.

The two met for an hour and a half the following evening in Macy's office. To his surprise, Macy told him that this would be the second-highest position in the entire space agency. As they talked, he asked why Webb would want him for the job, since they did not even know each other. Somewhat perplexed, Macy replied frankly that it was not Webb but the president whom he needed to impress, that this was, in fact, a presidential appointment. He now had Paine's full attention.

After the evening meeting, Macy called Webb at home and told him that he had "somebody who I'm sure is exactly the right man for you; I'm extremely impressed by him."[9] Webb, however, had never heard of Paine. He knew about TEMPO, had heard some good things about their work, and was familiar enough with GE to know that, being such a wide-ranging empire, it surely had many of the same management complexities as his agency. He was quite busy, however, and did not have time to meet. But since Paine was already in town and would be going back to California the next day, Webb told Macy to have him stop by NASA Headquarters the next day.

The next morning, he took a taxi to Federal Office Building Six. While waiting outside Webb's office, he struck up a conversation with Webb's secretary, who was impressed enough with his engaging nature to scribble on a notepad the words "The guy is alive!"[10] After a few minutes, Webb called him into his office. He was finalizing some notes for a hearing on Capitol Hill later that morning on the agency's budget for fiscal year 1968—a budget that included some rather significant cuts to the Apollo program. As the two began talking, it did not take long before Webb started rambling off names of people in Washington. What did Paine think of them? One was none other than Herb Hollomon, his old boss at General Electric from Schenectady. Hollomon had left GE in 1960 and was now the undersecretary at the Department of Commerce. The two quickly had a meeting of the minds and found that they liked and disliked the same things and people.

"Now, why do you want to do this?" Webb, with a stoic look, suddenly asked him. Paine had, in fact, recently received an offer from the GE corporate office that would have promoted him into a select pool of managers from which the company picked its vice presidents and top executives. But he had declined the offer. He was content and quite satisfied in Santa Barbara running what was essentially his own company in TEMPO. But the opportunity that Macy and Webb were now giving him, to enter public service at the national level, was, as he later recalled, very unexpected, different, and engaging. The presidential appointment, however, was the difference-maker.[11] They talked some more. By the time Webb had to leave for the hearing, Paine was very interested in the job, but said that he would only consider accepting if it was actually Webb himself (and not the president) who wanted him at NASA.

Webb was looking for a particular type of person to be his second-in-command. His top priority was to find someone who could act independently on his behalf when necessary. Paine met the other central ground rule that he had set: he wanted to bring someone in from the outside who had not been involved with the Apollo 1 accident. Once they met, he saw that Paine was engaging and articulate, and also had an impressively diverse portfolio of experience in the sciences, engineering, and management for someone his age (he was forty-six at the time). What is more, he came across to Webb as one who had the good judgment needed to make big decisions based on something other than political expediency.[12]

They shook hands as they walked out of the building. Before getting into his car, Webb told Paine to take some time and think it over. Although he and Macy were pretty sure that Lyndon Johnson would go along with whomever they recommended, they could not be certain. More importantly, they did not want Paine to change his mind and turn the president down if he were to get the offer from the White House. Paine was ready to say yes, but returned to Santa Barbara without any commitment. He waited exactly one week, then called Macy back and reaffirmed that he would accept the appointment if the president made the offer.

On the morning of Wednesday, January 31, 1968, Paine accompanied Webb to the White House to meet with Johnson. Moments earlier, as they walked by the Roosevelt Room to the Oval Office, Webb told him to just let the president do all the talking. Johnson, as it turned out, had been up most of the night. The military campaign in Vietnam that would turn into the infamous Tet Offensive by North Vietnam—a turning of the tide in the Vietnam War from which his presidency would not recover—had just started. The surprising and concerted counterattack by North Vietnamese and Viet Cong forces slowly but decisively swung the pendulum of public opinion in the United States strongly against the war.

Johnson had just returned to the Oval Office after a three-hour emergency breakfast meeting with various members of his cabinet and Congress. Halfway around the globe, the Viet Cong were still inside the US Embassy in Saigon. Paine recalled that Johnson looked haggard, his mind clearly not on the space program. He and Webb sat and listened as Johnson spoke quietly in a deep monologue on the situation of the country, the military reversal in Vietnam, and the great difficulty that his administration was facing. He needed good people to come into government service and help shoulder part of the burden, to serve him loyally, and to help him move the country through the trying time. It was with some effort that he slowly redirected his thoughts to talk about the space program. But as he did, his countenance brightened considerably.[13]

The meeting lasted only ten minutes.[14] Lyndon Johnson did not know Paine at all. During the 1964 presidential election, Paine had formed a local chapter of Scientists and Engineers for Johnson in Santa Barbara. He did this chiefly out of concern over Barry Goldwater's apparently cavalier stance on the use of nuclear weapons. The group was but a small, part-time operation in a little rented office. There, Paine and a dozen or so volunteers sent out letters on the weekends asking for donations on behalf of the Johnson-Humphrey ticket. His one and only foray into politics was very casual, with no direct ties to the national campaign. In fact, had it not been for the nuclear weapon issue, Paine would not have even bothered to oppose the conservative senator and Republican nominee from Arizona. Nominally a Democrat, he was not political. His party registration was usually whatever he thought would be in the best interest of the community where they lived at the time. When he and Barbara were in Massachusetts, he had registered as a Republican so he could vote for Eisenhower, mainly hoping that the new president might move the Republican Party toward the political center.[15]

It turned out that Johnson just wanted to meet him before appointing him. He went with Macy's recommendation and made Paine's appointment as the deputy administrator of NASA the next day with a brief mention in the daily White House press release. The US Senate easily confirmed him on February 7, 1968, and he was sworn into office on Monday, March 25.

The appointment drew little attention. Most in the space agency had not even heard of him. Being that he was such an unknown, reaction from the space community at large was subdued. The nature of the appointment did not garner much scrutiny either. Some national media outlets reported that there was some trepidation on the part of space program supporters on Capitol Hill. As an outsider and a Washington novice, would he prove adequate for the political tangos of the Beltway?[16]

Leaving Barbara and the family behind in Santa Barbara, he moved into a small apartment near the National Mall. He made the short walk to Federal Office Building Six every morning. At the time, NASA was using the seven-story building at 400 Maryland Avenue SW as its headquarters, sharing it with the Department of Health, Education, and Welfare. The $42,000 government civil service salary was a big step down from what he had been making at General Electric. He also liquidated the 770 shares of GE company stock that he owned in order to accept the nomination. To relax, he painted pictures of the city landscape on a small easel he had brought from California; most of the time, it just sat in a corner of the apartment. Barbara and the children (all but Greta, who was at the time away in college at Berkeley) soon joined him. When they did, they moved into an elegant old house on Ordway Street near the Naval Observatory.

On a personal level, Paine called the move to Washington a "recharging of his battery" after having spent nearly twenty years of his life working at General Electric. On the national level, getting Apollo ready to fly was NASA's top priority in the spring of 1968. A stack of Apollo 1 accident reports on his desk greeted him on his first day. Webb wanted him to review the management structure that was in place for human spaceflight to make sure that nothing was being overlooked. It had been twelve long months since the accident, and cleaning up the fallout was still consuming a great deal of Webb's time. To this end, he put Paine in charge of implementing the management changes that were coming down through the agency, from Headquarters in Washington to the various field centers around the country. Webb asked him to give the review his top priority and to wrap up these moves as quickly as possible.[17]

A consummate, powerful bureaucrat, Webb was a disciplinarian. For eight years, he ran the NASA Office of the Administrator with what he called "a pattern of operation through a series of grants of powers and authorities." Simply put, it was a way for him to delegate responsibility to his closest managers without losing control. Paine realized rather quickly that Webb was blessed with the uncanny ability to correctly judge people and give them just the right amount of authority.[18] On his first day, Webb gave him the final signature authority on all matters pertaining to human spaceflight. And when Webb was not around, he filled in as the agency's administrator at Headquarters. In this way, he had broad authority to make decisions on matters beyond human spaceflight. This extended to other programs, and, to a somewhat lesser extent, budget matters as well. In effect, the arrangement gave Webb and Paine reciprocal responsibilities atop NASA. In 1968, America was in the homestretch in the race against the clock and the Soviet Union to land a man on the moon. Critical managerial decisions were being made daily at the very highest level.

Paine was now one of the top three people in the agency. Webb was the administrator, and he was the deputy administrator. Homer E. Newell Jr., a highly-respected career mathematician and the agency's associate administrator, completed the triad. What Webb wanted was a formula at the top of NASA such that he, Paine, or Newell could give the final say on a matter with what he termed "concurrence after a limited review" by the other two. The arrangement prevented any one of them from being left out on any important matter. It also had the advantage of showing, on record, who was taking the final responsibility. In addition, he gave Paine a large amount of power by delegating authority to him on all things technical inside the agency. This gave Paine the final say from NASA Headquarters on all technical matters.

Paine once described his role as the deputy administrator of NASA in an interview. He called himself, and was soon known at Headquarters as "Mr. Inside." While he oversaw the technical execution of programs within the walls of NASA, Webb, as "Mr. Outside," did what he did best, which was to deal with Congress and the White House to secure a sound budget for the space agency and its plethora of programs. The arrangement worked, mainly because it catered to each man's strength. Webb had successfully used this shared approach to power with others for over thirty years to establish himself as one of the most dependable bureaucrats in Washington.[19]

Paine had to learn on the job, and quickly. He was not a career civil servant but he knew how things got done in the corporate world. At GE, he had watched how top managers with responsibilities on five continents ran a variety of businesses with varying degrees of success. How successful each was had much to do with how that person valued the individuals in the corporation. He recognized that a management system was, at its core, nothing more than a contract between managers and workers to follow through on a common goal; each individual had a role in the final equation. He preferred to let those under him assume their own styles and implement their own management methods. Colleagues recalled that his strength was the ability to manage people without their knowing that they were being managed.[20] He and Webb were, in this regard, very different. The different approaches that the two took to get things done were evident to those who saw them at work in Washington, and their contrasting styles often worked to their mutual advantage. He would later reflect that "it was remarkable how often we came out to the same answer, although through somewhat different paths."[21]

Soon after he arrived at NASA, he revealed what he thought the US should be doing in outer space. In the first speech he gave as deputy administrator that May, he proclaimed that America clearly needed to stand out among the nations as the world leader in science and technology, and "establish preeminence as a permanent space-faring nation." To make this possible, the United States had to select programs that deserved and would receive widespread national support. It was very important to have leaders who could motivate and guide the average engineer, technician, or secretary to attain goals all could understand. It was then up to NASA, and him, to make those goals attractive to the nation.[22]

He once called the NASA of the 1960s a unique sort of "management test bed." It required some unique and unprecedented organizational qualities for its time in order to try to accomplish something that had never in the history of civilization been done before. To their credit, leaders in Washington recognized this, for the most part. They gave the space agency far more freedom to experiment with new management approaches than other agencies. The leeway that was granted to NASA was important. How well NASA could marshal the best talent from American universities, industry, and the government largely determined how successful the country would be in space. But the mood of the country was a key factor; there was vocal skepticism of the space program, and it was not confined to inside the Beltway. Virtually no voice had been raised against the US space program ten years earlier, when NASA was in its infancy. Yet in 1968, after a string of remarkable achievements, many doubts were being raised.

The year 1968 epitomized the unrest of the "Generation of the Sixties." Social activism and the war in Southeast Asia dominated the news. The space program had to contend with those and other national concerns. Paine came to NASA because the agency was trying to meet a presidential mandate; time was now running short. In 1961, JFK, in a carefully orchestrated Cold War challenge, had called for America to land a man on the moon and return him safely to Earth before the end of the decade. The Apollo 1 accident had dealt this enterprise a sharp blow. NASA had not flown anyone in space since November of 1966. After Johnson became president following Kennedy's assassination, Webb found himself spending long hours trying to underscore to the new president, within the context of these other sobering national concerns, the importance of not losing the upper hand in space to the Soviet Union.

But Lyndon Johnson was, in fact, a political ally of NASA. As the ranking US Senator from Texas, he had been influential in passing the National Aeronautics and Space Act, which created the agency in October 1958. He had served as the chairman of the ad hoc committee that carried out the Space Act, and later chaired the Senate Committee on Aeronautical and Space Sciences. His clout had, in 1963, brought the prized Manned Spacecraft Center (MSC) and "Mission Control" to Houston—the word "Houston" soon becoming synonymous with the space age. (The center, which is actually on the outskirts of Houston in Clear Lake City, was renamed the Lyndon B. Johnson Space Center in 1973, following his passing.) As the vice president, he had chaired the National Aeronautics and Space Council, which oversaw the aerospace activities of the United States.

In his first full term as president, Johnson began pouring billions into domestic programs as he launched the Great Society campaign in an effort to combat poverty and social injustice at the height of the civil rights era. But by 1968, his administration was mired in the political and economic toll of these programs and the sharply escalating conflict in Southeast Asia. The space program no longer had the priority that it once had for the White House.

When Johnson appointed Tom Paine as the deputy administrator of NASA in February of that year, it was not with the thought that he would become the next administrator. Webb did mention the possibility to him when they first spoke, but only in passing and only in a perfunctory manner regarding the order of succession. Paine had dismissed it entirely. The scenario was also quite unlikely at the time, since it seemed very likely that President Johnson was going to be elected to a second term. This meant that Webb could have been expected to continue in his role.

But with the sharp decline in his popularity and his standing diminishing within his own party, Johnson unexpectedly announced (just two months after appointing Paine) that he would not run for a second term. The announcement shook Webb personally; he was blindsided. Paine recalled seeing his boss's attitude change almost immediately, as he began to noticeably loosen his grip and started transferring more and more signature authority to him. The pace picked up even more through the summer of 1968.[23]

With Johnson not seeking another term, a new president would occupy the Oval Office come January, and it now looked more and more like it was going to be a Republican. Webb was so political that he knew he could not survive such a change. He waited as long as he could, through the summer of 1968, before finally calling the White House to ask for a meeting with Johnson. "I thought it was important that there be an orderly transition from me to whomever the new administration decided to appoint to the post," said Webb.[24]

On the morning of Monday, September 16, Webb stopped by Paine's office just down the hall to tell him that he was heading over to the White House to discuss, among other things, a transition plan for NASA for when the president left office at the end of his term. It was not to tender his resignation; that transition would still be some way down the road. Paine recalled that Webb was very clear on that. Webb wanted to bring the subject up with Johnson mainly so the president could start thinking about how he wanted to position the NASA budget as he left office.

Over at the White House, however, as they talked in private, President Johnson took it differently. Within the hour, he had accepted Webb's "resignation" and called in his press secretary, George Christian, to have him announce to the White House press corps that the head of NASA was stepping down. Once Johnson made up his mind, it was too late for Webb to clarify his intentions. The White House announced that Webb had informed Johnson that he would resign from NASA on October 7, 1968, his sixty-second birthday. Paine recounted that he was certain this had not been Webb's intent; he believed that the private discussion took a turn such that "he and Mr. Johnson probably got somewhat swept away with their discussion and decided, well, this sounds like a very sensible thing to do, let's do it."[25]

Paine was the first at NASA to hear about it. Other than a short meeting with Homer Newell, he had not left his office all morning. At the White House, as Christian led Webb to the press conference (which Johnson did not attend), he managed to get a quick telephone call out to his deputy, frantically informing him of "a rapid development." The president would be asking him to temporarily assume the role of acting administrator. Webb never even had a chance to call his wife, Patsy, who heard the news through a neighbor who had heard a report on the radio. She quickly called Barbara.[26]

In his final press conference as the administrator of NASA, Webb tried to explain that by stepping down early he would give the agency the best chance to survive a change to a new president; someone else, who was not as political, could now be in charge. He continued, conceding finally, when pressed, that he had not been satisfied with the direction of NASA under his watch. Asked later what he meant, Webb admitted that he believed the space program under his watch had not been "able to achieve a first position in space" ahead of the Soviet Union.[27]

Paine, listening to the press conference, had not expected Webb to be so candid. He acknowledged later that Webb never gave him a full explanation of his resignation. "I believe that [Webb] felt that the best thing for NASA and himself, too, would be to put himself in the position of clearly resigning before the new administration came in," Paine speculated in hindsight.

> This would give him the opportunity to remove NASA somewhat from the political area, since James Webb, as the previous political appointee, would be out and the job of administrator would be open. It would give whatever administration that came in an opportunity, very calmly, to find the best man for the job without feeling a time pressure to remove Mr. Webb, which would be

> political, to assure continuity in the job, which [he] was most anxious to do. He also felt that this would give me the best opportunity to remain on as deputy, since I would have been in the deputy job for a little less than a year by the time the new administration came in, and by removing the top man, I would be given an opportunity to show what I could do with the agency.[28]

Johnson did not immediately name Paine as the acting administrator. He did not have to, however. Even with no presidential action, the deputy administrator, by law, assumed the role. By not stepping forward to name him acting administrator right away, the Johnson White House was, in effect, deferring the decision. Paine might still become the full-time administrator of NASA once a new president took office in January. But it would then be up to the new president, either a Democrat or a Republican, to decide whether or not he wanted to keep him.

The way the White House handled the Webb-to-Paine transition was the least risky move for Johnson. At the time, NASA had yet to launch a manned Apollo flight. Whether or not this was deliberate on the president's part, Paine said in hindsight, was unclear. But what was clear was that Lyndon Johnson ended up doing him a favor. The way he saw it, the nondecision by the White House took the pressure off him so that when the new president took office, he would not feel compelled to make a change just to replace someone from the previous administration.[29]

On a personal level, there was a huge difference between being second-in-command and being in charge. America's space program was his responsibility now. "God knows it is there, it gnaws at you all the time," he confided in an interview not long after. "There are nights when you don't sleep. I think there are very few jobs in the country—certainly none that I know of outside the military, and really not many there even—where a guy has to so put his neck on the guillotine, in the full view of the entire world, for the prestige of the country and the fates of some good people, and the professional qualifications of hundreds of thousands of people, all laid on the line."[30]

After Webb left, the White House essentially left Paine alone to do his job. He usually notified the president only after the fact when he acted on a given matter, not so much as a request for approval as to keep him informed. In the remainder of his term, Johnson did not overturn a single decision Paine made as the acting administrator. Years later he recalled that Lyndon Johnson was actually quite knowledgeable about the space program. Following the Apollo 7 mission in October of that year, the president would invite the three astronauts, Paine, and their families to his ranch outside Austin, Texas. After having lunch with Lady Bird Johnson and riding horseback around the ranch, they retired inside, where the five spent a leisurely afternoon talking about the space program. The high technical level at which they conversed delighted him. It was a side of Johnson that he had never seen before. He very much appreciated Lyndon Johnson and what he had done for him.[31]

6

GAINING SOME RESPECT

. . . the beginning of a movement that will never stop.
—Tom Paine on the eve of the Apollo 8 circumlunar flight

Had the United States not defeated the Soviet Union in the Moon Race by orbiting the moon in 1968 and then landing there the following year, history would likely know Tom Paine as the administrator who had the shortest tenure at NASA. The timing of the Apollo program, which landed twelve men on the moon, was his most fortuitous ally. He did not shape those events, but nor was he a mere minor player.

Four days after he took over as the acting administrator of NASA, Apollo 7 lifted off on a picture-perfect launch from Launch Complex 34 at Cape Canaveral.[1] It was the long-awaited maiden flight of America's brand-new spacecraft. Twenty-three arduous months had passed since the country last sent men into space. The cloud of the devastating Apollo 1 accident, though lightened somewhat, was still heavy, a not-so-subtle reminder to the nation of the pernicious nature of human space travel. The Apollo 7 launch was an emphatic response to the tragedy that had taken place on the very same launch pad just twenty months earlier. For the public, the maligned space agency now had the promising restart to the moon program that it had been waiting for. Astronauts Wally Schirra, Donn Eisele, and Walt Cunningham gave the program that shot in the arm as they orbited Earth 163 times over the next ten days.

NASA was now Paine's responsibility. Inside the hallways of the agency's headquarters, tension was high. As the one in charge, he had to be absolutely sure of one thing: could the Apollo spacecraft fly? NASA had no choice but to completely redesign the three-man capsule after the Apollo 1 fire.[2] Managers, engineers, and the astronauts who would have to fly it had made sure, to the extent that they could, that the new spacecraft was a much safer, more reliable vehicle than the original version that had claimed the lives of three of their colleagues.

From the outside, the Apollo spacecraft looked essentially the same. But the redesigned command-service module (CSM), which was known as the Block II CSM, was actually a brand-new spacecraft. It took the better part of two years, but NASA was able to rework every subsystem to improve safety and reliability. The entire electrical system of the command module—the part of the spacecraft that housed the three astronauts—had been rewired to reduce the number of wires and junctions that could serve as potential ignition points. Fireproof beta cloth, woven from Teflon-coated glass fibers, covered the interior. A single hatch that could open quickly in a matter of seconds from the inside now replaced the cumbersome and heavy, inward-opening two-piece unit that had previously trapped the crew. And perhaps most importantly, Apollo 7 was the first human flight of a Saturn-class launch vehicle. Although much smaller than the Saturn V that would later send astronauts to the moon, the Saturn IB rocket was still the most powerful ever launched with humans onboard.

It was just one flight, but from his vantage point in Washington, it felt bigger, much bigger. What the first successful manned mission of the program did was to put the nation's spotlight back on NASA. He was now confident that they had a spacecraft that might just actually be able to fly all the way to the moon and back. He called it the payoff for twenty months of recovery following an appalling setback, the lingering effects of which were still very much on the minds of his people throughout the agency.[3] Nowhere had the tension been higher than it was in Houston and at "the Cape" in Florida.

The staid mood in the days leading up to the launch had been noticeably more melancholy compared with the almost festive atmosphere that used to surround the Mercury and Gemini launches in the early days of the space program. This time around, Cocoa Beach on the Florida "space coast" had been much quieter. The mood had been pensive, as if those who were present expected that something bad was going to happen. A renewed sense of determination now prevailed among the agency's leadership. Acrimony within the ranks of NASA and toward its prime contractor, North American Aviation, began to fade. NASA was finally able to say it had passed a critical milestone in the Project Apollo timetable.

As he stepped up to the podium for his first official remarks as the head of the agency, Paine was cautiously sanguine. Although he would have liked to be more effusive, he reservedly called Apollo 7 simply "a very reassuring mission."[4] The fact was that his involvement on Apollo 7 had been minimal. He had taken over for Jim Webb less than a week before the launch. But the next flight would be quite different. The iconic decision to send Apollo 8 all the way to lunar orbit on Christmas 1968, in just the second manned flight of the program, turned out to be a watershed point in the nation's quest for the moon. Although he was still new to NASA, the decision won over his doubters. It began to ratify his position atop the agency in the eyes of his peers. In retrospect, it also exemplified the boldness in decision making that was a hallmark of the American space program in the 1960s.

In September of 1967, nearly eight months after the Apollo 1 accident, George Low and Owen Maynard from the Apollo Spacecraft Program Office in Houston had briefed George Mueller at the Office of Manned Space Flight in Washington on a milepost schedule for Project Apollo going forward. Houston had come up with a progression of seven missions that, if successful, could land a man on the moon by the end of 1969. Maynard had assigned a letter to each of those missions:

A and B. Unmanned flights of the Apollo spacecraft
C. Manned flight of the CSM in Earth orbit
D. Manned flight of the CSM and LM in Earth orbit
E. Manned flight of the CSM and LM in high-altitude Earth orbit
F. Manned flight of the CSM and LM in lunar orbit
G. Manned lunar landing[5]

Since they were unmanned, the A and B missions had already flown in 1966 without much publicity. Apollo 7 then completed the C mission. The next step was to add the lunar module and test the complete Apollo spacecraft in Earth orbit. There was only one problem. The LM—that critical piece of hardware needed for the D mission—was significantly behind schedule. Mueller had wanted to fly the D mission by the end of 1968, but by June of that year, it was already apparent that the spacecraft would not be ready to fly until the following February at the earliest.

The entire timeline was critical. With the end of 1968 approaching, NASA had less than eighteen months to work with. On top of that was the critical unknown of how far along the Soviet lunar program really was. The CIA kept Paine informed on intelligence, but even they could not be certain. The Soviet Union had suffered a major setback when Chief Designer Sergei Korolev died unexpectedly in January 1966 following surgery. But he had already set into motion the development of a massive launch vehicle (publicly revealed by the CIA only after the fall of the Soviet Union in 1989 to be the N1, a 344-foot, five-stage rocket more powerful than even the American Saturn V) designed to send two cosmonauts to the moon in their new Soyuz spacecraft. The US knew only that the first flight of the Soyuz in the spring of 1967 had failed miserably. Veteran cosmonaut Vladimir Komarov had been killed on reentry. But the CIA could not be sure just how much ground they had actually lost.

Knowing that the lunar module would not be ready in time, Mueller wanted to use the next mission to further prove the flight worthiness of the CSM by flying it in a high-altitude, high eccentricity elliptical orbit. The farthest point on such a trajectory would take the spacecraft to the vicinity of the moon, where its response to a deep space environment could be tested, but it would remain in Earth orbit should an issue arise. In other words, they would fly the E mission but without the LM. This was a conservative but not very engaging option—essentially an elaborate repeat of Apollo 7. Even if nothing went wrong, it would not accelerate the overall landing schedule by much.

George Low, now running the Apollo Spacecraft Office in Houston, then put forth a captivating and rather daring idea: since Apollo 8 was going to fly away from Earth, why not send it all the way to lunar orbit? A December launch window (as dictated by orbit mechanics and the relative position of the Earth and moon) presented such an opportunity. If NASA could pull it off, the "C-prime Mission," as he called it, could shave six months off the schedule. Moreover, if successful, the agency could try to put a man on the moon as early as the summer of 1969.

For the plan to work, Low and the team of Apollo managers from the Manned Spacecraft Center in Houston, the Marshall Space Flight Center in Huntsville, Alabama, and the Kennedy Space Center in Florida had to persuade three people: Webb, the NASA administrator; Paine, his deputy; and Mueller, the associate administrator for manned spaceflight.

Of the three, Webb, who had been on uneven terms with the Apollo managers even before the fire, would be the toughest to convince. Mueller, who had approved the original order of the Apollo missions, would be difficult as well. The advantage in dealing with Mueller, however, was that he still had Webb's trust, for the time being. Furthermore, he worked for Paine. Thus, in the eyes of the Apollo managers, Paine was the focal point. Convince him and the plan had a chance.[6]

On Wednesday, August 14, 1968, they got their opportunity. Webb and Mueller were away in Vienna, Austria, attending a United Nations conference on the peaceful uses of outer space. The largest personalities of the US space program—none more so than Wernher von Braun from Huntsville, already a world figure—gathered in the conference room next to Paine's office at NASA Headquarters. Low got up first. He began by briefing the group on technical issues of spacecraft readiness, the status of the communications network, crew readiness, and ground support requirements. Next came a discussion of program risk and schedule.

Quietly Paine listened, speaking briefly now and then only when he had a question. For the next three hours, the men who ran Apollo made their case. One by one, each went to the front of the room and showed with one viewgraph after another the technical and programmatic reasons as to why Apollo 8 should take the gamble and fly to the moon before the end of 1968.

Understanding that it was a gamble, he finally looked at von Braun and reminded him that a crew had not yet even flown on a Saturn V and "now you want to up the ante." When everyone had finished talking, he went around the table:

> "Once you decided to man 503 [the serial number of the Saturn V slated for Apollo 8], it did not matter how far you went." (Wernher von Braun)

> "It is the only chance to get to the Moon before the end of 1969." (Deke Slayton)
>
> "I have no technical reservations." (Kurt Debus)
>
> "It will be a shot in the arm for manned spaceflight." (Julian Bowman)
>
> "The plan has my whole hearted endorsement." (William Schneider)
>
> "There is always risk, but this is a path of less risk. In fact, the minimum risk of all Apollo plans." (Robert Gilruth)

Chris Kraft, director of flight operations in Houston, was the most cautious. "Flight Operations will have a difficult job here; we need all kinds of priorities. It will not be easy to do, but I have confidence." George Low, sitting next to Paine, turned and simply said, "Assuming Apollo 7 is a success, there is no other choice."[7]

Paine had to consider the alternative, which was to not fly the mission until the next launch window came up. The timing was critical. If they did not take advantage of the December launch window, the mission would have to wait until late the following spring at the earliest. It would mean a delay of close to six months.

He was new to Washington and new to NASA. In the six months that he had been with the space agency, he had managed to develop a good rapport with his closest advisers in the Office of the Administrator. This was, however, his first exposure to the entire team of Apollo managers from across the country. Sitting at the head of the large conference table, he found their confidence convincing, but was not at all sure that they were telling him everything. He later recalled that the sheer audacity of their action spoke for itself. It surprised and impressed him at the same time.

He remembered von Braun as the most vocal. The enigmatically persuasive rocket engineer, along with Eberhard Rees, deputy director of the Marshall Space Flight Center, presented him with a very strong case that the Saturn V was now at the point where the next launch could take the risk of flying with men on board. They were most confident, and they needed to be. He had made it a point to visit the pair of famed German rocket scientists in Alabama as one of the very first things that he did after joining NASA. He respected von Braun's considerable reputation, and considered him something of a fellow visionary, but with more flair and status. Rees he looked to more as a graybeard adviser—the top-notch engineer of the World War II V-2 terror weapon, with whom he could speak on a very technical level concerning all aspects of the Saturn V. As the meeting adjourned, he said to no one in particular, "We'll have a hell of a time selling it to Mueller and Webb."[8]

Foremost on his mind was weighing the known issues against the unknown risks. A risk is a potential issue waiting to happen, and there would be plenty in Low's proposal. He had to consider his next steps carefully. He did not want to risk breaching Webb's trust. One option was to just wait and bring up the idea with Webb and Mueller completely off the record—talk it over with them after they had returned from their trip and see what they thought. He was, however, now also in a position where he could change the course of the US moon program. From a technical standpoint, no other option would accelerate the program in this way. By seizing the opportunity, he could elect to take the riskier path that, if successful, would provide the chance to get the US to the moon faster.

But he did not have the authority to make the decision unilaterally. Webb and Mueller would not be back in the States for another week. That was too long. He picked up the phone and called Mueller at the embassy in Vienna. Mueller reacted with disbelief. He hated the proposal. Mueller had arrived at NASA five years before Paine, but now found himself working for the new guy. He had a recalcitrant nature, one that was well known inside the hallways of NASA, something he often used to his advantage. Relying on the force of his personality, he usually got his way. Most of the time, he turned out to be right.

After he hung up the phone, Paine picked it up again. This time he called Webb. He was not as confident now, after the conversation he just had with Mueller. He would later reflect on that phone call. "I can imagine how he felt, getting a lunatic telephone call at the American Embassy there from a fellow who had been in the agency for all of six months, telling him we were going to go around the Moon."[9]

Indeed, Webb reacted even more strongly than Mueller. There had been talk of schedule changes before, but none had the ramifications of this. He was furious. No way was he again going to be put into a position where he would have to defend the agency should another accident happen. He raised all sorts of objections: that Christmas was no time to send astronauts a quarter of a million miles from home, and that people were just not going to be receptive to such a radical idea. "I've explained all this to you!" a stunned Webb told him.

It was clear to Paine that NASA could not afford to simply dismiss the idea of orbiting the moon on Apollo 8. The space agency had to seriously consider it; too much was at stake. He tried to convey that to Webb without going into all the technical details. (He later recalled that it was difficult to talk since they were sure that the Soviet Union was monitoring their call.)[10]

The next day, he, Sam Phillips (Apollo program director at NASA Headquarters), Willis Shapley (associate deputy administrator), and Julian Scheer (the NASA Public Affairs Officer) met in his office. By the end of the morning, they had drafted a seven-page internal memorandum that explained his case. He had it sent to Webb using secure State Department encrypted cable that afternoon. In it, he provided details on the potential impact on the remaining Apollo schedule of changing Apollo 8 to a circumlunar mission, including flight plan options, hardware readiness, and the implication for future crew assignments. To make things easier for Webb, Scheer included a draft statement for a possible press release.

In Vienna, Webb got the memo and conferred with Mueller. Now somewhat calmer, he still questioned the pace at which this was all taking place. He dictated a response and cabled Paine back that studies should first be carried out and plans prepared so that some flexibility might be injected into the final mission decision. Webb told him that he still had many questions that would have to wait until he got back to Washington.

Paine sent more secure messages and seemed to have at least convinced his boss that the option of a lunar orbit should be kept open. After several long-distance cable exchanges across the Atlantic, Webb tentatively directed him to notify the White House that NASA was now thinking about changing the mission plan but had not yet made a decision.

Arriving in his office the next morning, he put exactly that in writing and had it sent over to the White House. By doing so, he had effectively raised the bar and White House expectations such that NASA could not turn back.[11]

Paine later admitted that even as he lobbied for Webb's endorsement, he had to also think that maybe the managers of the Apollo program were manipulating him from inside the agency. The timing of the meeting had been suspicious. He knew of the tension between Gilruth in Houston and Webb in Washington. With their "opposition," Webb, out of the country, Houston might very well have capitalized on the opportunity to make their move. Never fully convinced otherwise, Paine remained skeptical. Even after Apollo 8 had flown successfully, he still thought this might have very well been the case.[12]

He did not know Gilruth and the people in Houston as well as Mueller and Phillips did. But he knew them well enough. On the decision about whether to go ahead with the mission, he at least trusted their judgment. Gilruth, the fifty-five-year-old aeronautical engineer who had worked his way up in the space agency to become the director of the MSC, had impressed him by how he had held the MSC accountable for its actions after the Apollo 1 fire. As a result, the space center was able to enact the right recovery actions from that tragedy. Webb, on the other hand, had lost all faith in Gilruth by this time. After the accident, he spoke to Gilruth mostly through Phillips or his deputy George Hage.

Paine wanted to see Apollo 8 go to the moon, and knew that the timing was right, regardless of Webb's objection. The clash between the top two people at NASA over the flight of Apollo 8 soon ceased to be private. He kept the pressure on his boss regarding the decision, and interpreted Webb's vague response to his proposal as a de facto okay. NASA's announcement to the news media on August 19 still (purposely) de-emphasized the lunar option, as Webb wanted. The embattled NASA administrator was by now very cautious in his every move with the Johnson White House. Two days earlier, Phillips and Hage had met with Gilruth in Houston to give MSC the go-ahead to replace James McDivitt's crew with Frank Borman's crew.[13] Once that was announced to the public, the decision for Apollo 8 to orbit the moon became a fait accompli. NASA could not turn back.

Webb knew that another catastrophic accident on the scale of Apollo 1 would mean the end for him at the agency. It would also, in all likelihood, collapse the empire that he had built. He thought it was a bad decision—even after the mission turned out to be an unparalleled success a few months later. He questioned just where disagreement stopped and insubordination began. Webb and Paine remained close, but on the Apollo 8 decision, they fought.

Paine had no regrets. He later compared Webb's inaction during that critical juncture of Project Apollo to the similarly strange actions of Admiral Horatio Nelson. The British flag officer from the history books of the American War of Independence had lost his right eye fighting the French:

> I remember[ed] it was a case of Nelson's holding the telescope to his blind eye. I felt it was better to preserve the option to fly C-prime lunar orbital. I was protecting the policy decision when all the technical people felt that lunar orbit was better for the whole program. If the Soviets flew their lunar orbit or circumlunar flight in January [1969], and we hadn't done anything about ours, Jim Webb would have been attacked for turning down our opportunity. I did it to protect him. . . . Furthermore, there was no way that [NASA] could secretly take such actions.[14]

The decision was risky. It was risky for Paine on a personal level and risky for NASA as an organization. If anything went wrong, the political and public repercussions would be most severe, perhaps even irreparable. His foremost technical concern was with the spacecraft's main engine—the large rocket engine at the back of the service module called the service propulsion system. A failure would strand the astronauts in lunar orbit. In October, he again asked Mueller for the engineering qualification numbers. Mueller went over the test history with him: only four in 3,200 firings had failed, dating back to the early prototypes. Most importantly, said Mueller, the engine had never failed to fire in its final production design version.[15]

As December approached, Paine talked again with Rees and von Braun. Of everyone involved in the discussion, it was they who had sold him on the mission, but they had never downplayed its difficulty.[16] Apollo 8 would be highly ambitious on a number of levels. It would be the first mission in which all program elements had to function together. It would also be the first human flight of the Saturn V launch vehicle. Here on Earth, it would be the first "all-up" test of the Manned Space Flight Network that would track and communicate with the spacecraft all the way to the moon and back. The mission would be the first exposure of an Apollo spacecraft to the deep space environment. Ambitious, yes, but if all went well, six months could be saved.

On Thursday, November 7, Mueller convened an internal NASA certification board review. They concluded that all systems were go for a December launch. Another review three days later by the team of industry contractors reached the same conclusion. The next day (Veterans Day 1968), Paine held the final go/no-go meeting in Washington. He had earlier asked Mueller to brief the mission plan to Donald Hornig, science advisor to President Johnson, since Mueller and Hornig knew each other. Mueller told him that the White House was on board with the launch. Gerald Truszynski, NASA's associate administrator in charge of the tracking networks, then got up and presented a short briefing that declared the Manned Space Flight Network ready. Lieutenant General Vincent Huston reported that the Defense Department was ready to support recovery operations in the Pacific. One by one, Phillips, Low, Kraft, and Rocco Petrone, director of launch operations at the Cape, all reiterated their support. The three-hour meeting raised no new concerns for Paine.

After all the others had left, he walked back to his office, picked up the phone, and called mission commander Frank Borman. The crew was "100 percent go," Borman said without a pause. That was the final confirmation—Paine just needed to hear it for himself from the person who was going to command the mission. He hung up the phone, read the letter of record over one more time, took out his pen, signed above his name, and approved the mission.[17]

As soon as he signed the paper, he picked up the telephone and called President Johnson's chief of staff, James R. Jones, with the news. Jones walked the few steps to the Oval Office and quietly put a note on Johnson's desk. By happenstance, Lyndon Johnson was meeting with President-elect Richard Nixon at that same moment. Johnson paused, read the note, said not a thing, and showed it to Nixon. Although presidential historians would chide Johnson in the years to come as one who mismanaged by micromanaging the Vietnam War from the Oval Office, in the space arena, "he never questioned our decisions," according to Paine. "We let him know after the fact and that was that."[18]

Paine later revealed to journalist Robert Sherrod that he was certain the decision would engender political accusations. Opponents suggested that it was a deliberate move by NASA to present a space spectacle to send an unpopular president out on a high note. He was ready for those attacks and was prepared to defend the space agency. The Johnson administration was indeed in a political mess by the summer of 1968, plagued by a plethora of domestic and foreign problems. But no pressure of any sort ever came from the White House. In fact, Paine did not have to use his prepared statements because negative publicity never materialized, as the moon mission captured the imagination of the public worldwide.

The year 1968 had been a long one for the space agency. But with Apollo 8, NASA took a gamble and it paid off. Paine was convinced that any ploy by critics of NASA on Capitol Hill or in the media would have only ended up backfiring had there been an accident on the flight. He remembered the flight as a very good thing for the country that year: "History will record that I think anybody would have done what we did. It was the sensible and right thing to do, and of course, the success of the mission vindicated it. But it was in the back of my mind that there would be people who might allege politics here." He was a scientifically minded man, an engineer. He thought in technical terms, and his decision was a technical decision. He considered the politics, of course, but only after the technical justification had allowed it to be germane. Inside NASA, people knew that Apollo 8 had bought valuable time and maximized the chance for a landing attempt the following year. His dear friend Robert Seamans went one step further and called the Apollo 8 decision *the* tough decision of the Apollo program.[19]

The stark, desolate face of the moon astounded the first human beings who saw it from lunar orbit. As Apollo 8 circled the moon on Christmas Eve 1968, astronaut Bill Anders described its dull gray, cratered surface as "a sandbox all torn up by children at play."[20] In an unscripted moment on live television, the crew took turns reading the Creation account from the book of Genesis.

Paine thought primarily of the Apollo 8 crew, and of Frank Borman in particular. He would call Borman an invaluable asset to the American space program after seeing his performance before a joint session of Congress after the mission. He was impressed enough to recommend Borman to Vice President-elect Spiro Agnew as his first choice to be the executive secretary of the Space Council in the new Nixon-Agnew administration.[21] When Borman turned that down, Paine asked him to be his liaison to Congress and a personal emissary to NASA's international partners.

The wholly unpredicted, worldwide outpouring of support for Apollo 8 confirmed Paine's belief in American exceptionalism. He encouraged the flag-waving. "People everywhere have taken this not as an American triumph, but merely that America was the very affluent, wealthy, capable country that had the will and the resolve to carry this Apollo program out, [and] that really we were doing this for all mankind and in the name of all mankind."[22] The payoff was about more than American dollars and cents. He called the mission "poetic," an uplifting Christmas present that the space-faring pioneers of America gave to the world that turbulent year. Most gratifying to him was the reaction from the Soviet Union. Ten cosmonauts sent congratulatory telegrams. Many leading Soviet scientists praised the accomplishment. Moscow, in an uncharacteristically conciliatory move, called the flight "a feat to be remembered for the ages."[23] It gave the world a brief respite from the images of Vietnam, the Tet Offensive, the My Lai massacre, the ugly urban riots, the shocking assassinations of Martin Luther King Jr. and Robert Kennedy, the Soviet invasion of Czechoslovakia, and the many other tired headlines of 1968.

Overnight, Apollo 8 vindicated NASA in the eyes of the public. Paine praised the mission as one that proved "we did know what we were doing; we were able to carry out the commitments we had made to the American people."[24] In his words,

> In the cynicism and disarray that had followed the assassinations, the cities, student unrest, Vietnam, the convention, and

> all the rest, I think that the country had lost sight of the fact that meanwhile, out in Houston and down at Cape Kennedy and in Southern California, the space agency was steadily moving ahead. It came almost as a surprise to people that all of a sudden, after not flying men for a long time, we were suddenly able to carry out this very advanced mission. It was not a surprise to us. We had been working on it.[25]

His defining moment at the agency came just six months after President Johnson appointed him as the deputy administrator, a fact that did not go unnoticed by space insiders. Looking back, Chris Kraft recalled that, from his vantage point in Houston, it suddenly became simpler to get things done through Washington. "We knew we had a friend in Paine."[26]

7

I'M NOT A POLITICIAN

Truth is putting together a system in such a way that you can fly to the Moon and back.
—Tom Paine, replying to a question on the meaning of being able to go to the Moon in 1969

As the calendar turned to 1969, NASA's big concern was with the Apollo lunar module. But Tom Paine had a bigger problem. On November 5, 1968, Richard Nixon defeated Hubert Humphrey in an election remembered for antiwar protests, riots, and the assassination of Senator Robert Kennedy. As Nixon was sworn in as the thirty-seventh president of the United States on January 20, Paine had been on the job as the acting administrator of NASA for just over three months. All indications coming out of the White House pointed to the new president as being noncommittal about Paine, at best.

By February, Nixon was being told by his advisers, and by science advisor Lee A. DuBridge in particular, that he must make a decision. Did he want to keep Paine or not? White House Communications Director Herbert G. Klein advised him that an official appointment had to be made very soon. Public opinion was still high, but the president should not risk jeopardizing the possibility of a moon landing by the year's end.

Congressional proponents of NASA wanted the situation settled as well. Like other agency heads from the Johnson administration, Paine had tendered his resignation on Inauguration Day as a formality. But unlike the others, his had not been accepted. The NASA post was, in fact, the last major appointment that Nixon still had to make. It was a critical selection, as it now appeared that he, a Republican, would, with no small irony, wear the laurels of a moon landing after eight years of Democratic leadership. Richard Nixon had always associated the space program with John Kennedy, his late archrival, and had made some very negative statements about it in the past. But in early 1969 it appeared to Nixon that if Apollo were to fail on his watch, it could be the one thing that might scar his presidency.

Paine was far from Nixon's first choice for the job.[1] Shortly after the election, the Nixon transition team had concluded, to no one's surprise, that a Republican should lead the space agency. Washington insiders quickly reported that Simon Ramo had turned down the job in January, even before Nixon took office. Ramo was an impressive electrical engineer with a long resume in the development of the nation's ballistic missile programs. By the 1960s, he had become a renowned aerospace leader and successful business executive. Years earlier, he had cofounded the aerospace giant TRW (today part of Northrop Grumman) in southern California; its sales had included over $50 million (about half a billion in 2017 dollars) of business with NASA in just the previous year. He had also led the development of the Atlas rocket. NASA had used it on Project Mercury to send four of its astronauts into orbit. Ramo was also well known in Washington, having previously worked in the Eisenhower administration. But Ramo's interest was now clearly focused on his lucrative *Fortune*'s Global 500 company.

Ramo's was not the only name on Nixon's short list. Patrick E. Haggerty, chairman of Texas Instruments, Hilliard W. Page, general manager of the General Electric Valley Forge Technology Center, L. Eugene Root, a top executive with the Lockheed Missiles and Space Company, and retired Air Force General Bernard A. Schriever were also offered the position. All turned Nixon down.[2]

Schriever was a particularly intriguing choice. The former ballistic missile chief of the United States Air Force was an ardent Nixon supporter. He was very well qualified, and was held in the highest esteem by Republicans and Democrats alike. But he too had just started his own company after thirty-five years in the military. The venture was growing and becoming quite successful. Only recently, he had opened offices on both coasts. He told the White House that he simply had too many personal commitments to consider taking the job.[3]

Throughout January and into February, Paine purposely kept a low profile and did not publicly campaign for the job. He had enough worries. But privately, he told his closest colleagues that he was convinced Nixon would end up sending him back to the job of deputy administrator. This came after the *Washington Star* reported that he was not even considered a candidate for the job by the new White House.[4] The mood on the top floor of Federal Office Building 6 was tense. He was prepared to go back to being the deputy, but only as a last resort. "My job," he said with more than a hint of frustration in his voice the first week of February, "is to give President Nixon everything I can; nothing more."[5] He too wanted a decision from the president, but his sources were coming up empty. The White House continued to give out no hint as to whether or not they would keep him. "Nixon seems to be having a hard time getting people to take high-level jobs. . . . If he waits too long, Homer Newell can take over the agency."[6]

With five Republicans having turned down the job, the White House took another look at Paine. By now, sentiment was mounting inside Washington to keep him on. It might, in fact, be the only workable option that Nixon had. Staying with Paine had many advantages. First, all of the directors at the NASA field centers across the country already knew him. One of the first things he did after coming to NASA was to tour every field center and report back to Webb on their status. The wisdom of bringing in a new person to replace someone who was already familiar with the agency and already known by the members of Congress who worked with NASA thus became less clear. Paine was also as politically neutral as someone in his position could be. This helped him. Members on both sides of the political aisle recognized that as an asset. A strong endorsement came from Nixon's acquaintance from California, Nobel Laureate Charles H. Townes (who was, by coincidence, on the NASA Science Advisory Committee). Frank Borman, an outspoken Republican, also endorsed him. Even Lee DuBridge, initially lukewarm to the idea, changed his mind about Paine. This support, plus the need to maintain the momentum gained from the recent success of Apollo 8, pointed to Paine keeping the job.

The final recommendation came from Bob Seamans, who told Nixon's newly appointed defense secretary, Melvin R. Laird, "I can give you a very straightforward, simple answer, Mel. Ask the president if he wants to carry out the lunar landing this year. If he does, make Tom Paine the administrator. But if he wants to run the risk of not going this year, then bring in somebody else."[7] By the last week of February, Nixon was convinced that he was out of options. His chief of staff, H. R. (Bob) Haldeman, asked Deputy Assistant Alexander Butterfield to move forward on Paine and to do the necessary background checks.[8]

On March 5, 1969, as Apollo 9 circled Earth, the president announced at a West Wing ceremony presenting the Robert H. Goddard Memorial Trophy to James Lovell that Paine would continue in his position as the head of NASA.[9] "We found, as is often the case, that the best man in the country was in the program."[10]

"I am very pleased and very touched by your confidence in me," Paine responded with great relief as he smiled and looked at Nixon. "I believe in the space program. I believe in this country, and I think that this country should indeed be the preeminent nation in spacefaring, and with your help, Mr. President, I am sure that we can go ahead in the next four and eight years to really see that the NASA program in the second decade of space will really outperform the accomplishments of the first."[11]

The result was the closest thing to a unanimous commendation for a presidential appointment in Washington. Nixon was happy and so was the Senate. It easily confirmed Paine on March 20, and on Thursday, April 3, Barbara, George, Judy, and Frank Paine watched (eldest daughter Greta was away in college) as Vice President Spiro Agnew swore him in in a small ceremony at the Old Executive Office Building.[12]

With his appointment now official, he had to find someone to be the deputy administrator. Paine had his own short list of names. Raymond L. Bisplinghoff was at the head of that list. The aeronautical engineer had previously served NASA as an adviser and assistant administrator to James Webb. Bisplinghoff had, however, just accepted an appointment to be dean of the Schools of Engineering at MIT, and told Paine that he was unavailable. Sam Phillips, the Apollo program director at NASA Headquarters, was another person he considered but did not nominate. Paine thought that it was best not to recommend him in order to maintain a clear line between military and civil space. Phillips was a career military officer. His duty assignment with NASA was winding down and he would soon be going back to the Air Force. Physicist Charles Townes, whom Paine knew when they were both in California, was another. Townes would have brought considerable credibility to the agency in the areas of advanced research and science programs, but he was unavailable, being too entrenched in his position at the University of California to consider leaving.

Several factors were figuring in his decision. Foremost was that he needed to choose someone who could run the agency in his absence. He also wanted a person with a strong technical background, who could make technical judgments in a variety of areas. The person should not be afraid to call things right or wrong as he saw fit. Since the position would be a presidential appointment, a known Republican would be best. That would provide the White House with an advantage if he were to leave the agency before Nixon's term was over. Finally, he wanted someone of relative youth, in his forties, whose reputation had not yet been cemented inside the Washington establishment.[13]

At the top of his list was Frank Borman. The former Air Force colonel had recently announced his retirement from active astronaut status following Apollo 8. Borman had all the qualities Paine was looking for. Besides the obvious technical qualifications, he had the right connections and was very close to the new Republican White House. Paine, in fact, thought that Frank Borman would eventually have made an excellent administrator if he had chosen to stay on with NASA. But the veteran astronaut had made it widely known that for family reasons he wanted to explore something besides the moon and try his hand in the lucrative world of corporate business. (He became the chairman of Eastern Airlines shortly after leaving NASA.) Despite having seriously considered him, Paine never got the chance to make Borman the offer.[14]

Time was running out. He turned his attention to selecting someone from inside the agency. The person who had led the charge on Apollo 8 the year before was George Low. Paine trusted Low. Low was a career NASA man. He had joined the staff of its predecessor agency, the National

Advisory Committee for Aeronautics, back in 1949 as an aeronautical engineer. While his technical specialty was in the area of fluid dynamics, his greatest contribution came in the area of human spaceflight. Low was a central figure in Projects Mercury, Gemini, and Apollo—NASA's three iconic manned spaceflight projects of the 1960s and early 1970s. After Project Mercury had successfully put the first Americans into space, Low transferred to Houston to be the deputy center director of the MSC. After the Apollo 1 fire, Webb put him in charge of the Apollo Spacecraft Program Office. Working closely with senior-level managers and North American Aviation, he successfully directed the redesign of the Apollo command-service module.

In Low, Paine found someone who understood the nuances of the agency and how things got done. He was already well known at NASA Headquarters, was easy to talk to, and had established an impeccable reputation with people at all levels of the Manned Spacecraft Center, from draftsmen to engineering managers to center director Gilruth. Low was one with whom he could brainstorm technical decisions about the systems engineering side of Apollo at a very foundational level, something that he could do only in an uncomfortable way with Mueller. After serving as Paine's deputy, Low would go on to be a key figure in the development of Skylab and the space shuttle. He would leave NASA in 1976 for the Rensselaer Polytechnic Institute, America's oldest technical research university in Troy, New York. There he would stay as the president of RPI until his death in 1984.

Paine immediately saw Low's impact at NASA Headquarters. Through the spring and into the summer of 1969, Low would continuously identify and bring in people from other parts of the agency. Most were young managers with solid technical backgrounds—rising stars in the middle ranks of the organization. He wanted to inject a fresh approach into the top tiers of NASA with these younger managers. Low had a plan. After they had been in Washington a few years, he sent them into different parts of the agency. They took with them a different outlook based on what they had seen and learned at Headquarters. Some prominent people in the space community whom Low recruited to Washington included Gerald D. Griffin and Joseph P. Allen. Griffin was an Apollo flight director who in 1982 became the director of the Johnson Space Center in Houston. Allen was a scientist-astronaut and an assistant administrator. In 1978, he returned to Houston and flew as a mission specialist on space shuttle flights STS-5 (1982) and STS-51-A (1984).[15]

The day he was nominated by Nixon, Paine received a call from DuBridge over at the Eisenhower Executive Office Building. The president wanted to see the NASA budget. Certainly Paine's greatest challenge during this time was the constant need to defend the agency's budget. Fighting bruising budget battles with the White House and Congress was just not Tom Paine's forte. His first battles had come with the agency's interim operating plan for fiscal year 1969, when LBJ was still the president. NASA's annual funding was then just under $4 billion, slightly less than what it had been the year before. President Johnson had eliminated $375 million right before leaving office, and another $350 million had been deducted by law as a result of the Revenue and Expenditure Control Act of 1968. Dollars were being cut from every program; manned spaceflight was not spared. This was before a single Apollo mission had left the launch pad.[16]

Even when he was the acting administrator, he had devoted a considerable amount of time to long- and short-term planning. He wanted to formulate a space program that he could sell with a reasonable degree of success to the Nixon White House and the 91st Congress. Both had a tight grip on the agency's future by controlling the level of funding the agency would see once the first manned moon landing had been achieved. To this end, it was his responsibility to make sure that Nixon did not make further cuts on top of those that had already been made by Johnson.[17]

"I often wonder what my opposite number, Academician Anatoly Y. Blagonravov, does when he goes to the Kremlin for money," he said in an interview that summer. "He proceeds in secrecy, while we are in a fishbowl here. I would like to know what arguments are used, how the power is disposed. For one thing, he is dealing by and large with people who have technical degrees. I confront law school graduates. I find that the members of the House committee are highly individualistic [and] answer only to their constituencies. You get tart, tough questions."[18] He told

Congress that NASA needed $5 to $6 billion to successfully implement all its plans. Even he conceded, however, that this would not be possible as long as the Vietnam War continued.[19] Congress was holding the purse strings very tightly, and Paine was finding it very difficult to convince them to loosen their grip.

In February, he talked at length with Clinton P. Anderson, chairman of the Senate Committee on Aeronautical and Space Sciences. The conversation eventually turned to the intrinsic value of the American space program. Anderson liked NASA, but did not like the question that was so often asked of him, "What has NASA done for me?" He asked Paine that very same question.

Paine told him to think in terms of technology transfer; that is, spinoffs produced by the space program. He did not mean just new hardware or new machines. The greatest spinoff from space exploration was the transference of knowledge. This included all of NASA's programs and efforts that moved new information and technology to the public. Once that had been done, he continued, it was up to the American public to see the knowledge put to good use. In other words, space exploration was only in part about outer space. He urged the senator to appreciate space exploration for its underlying benefits. Many technologies from the space program had been used to advance American wealth and power. While they might have been developed in time without a space program, the process was at least accelerated. New discoveries in fields such as material science, medicine, electronics, and safety were made because the space program came along when it did.[20]

Shortly after his appointment, Paine told an audience that the nation was right to expect that NASA undertake a bold initiative like going to the moon. He believed that his agency was one of the few large government organizations left that still encouraged independent thinking. To him, it was the people inside the space program—the gifted scientist, the brilliant professor, the creative innovator, the strong manager—who made NASA excellent in the 1960s. In a 1967 essay, he wrote that the greatest single achievement of the space age might very well have been the formation of NASA. It allowed the energy, talent, and intellect of America to be unleashed in a clearly focused and very public way.[21]

He had a simple, pragmatic management style. As the head of the nation's civilian space program, it was his job, first and foremost, to define and articulate goals and to sell those programs to the American public. It was, in essence, to justify the existence of a space program. One of his biggest regrets, he later reflected, was his inability throughout his time at NASA to get across to the American people a satisfactory answer to the seemingly simple question: "Why shouldn't we take the money we are spending on the space program and put it into the cities or the poverty program?" The answer, he maintained, was not quite as simple as the question.[22]

Paine wanted a permanent human presence in space beyond Project Apollo. It did not have to stop with the moon. He recalled on numerous occasions that it could be done as long as NASA selected programs that deserved and received widespread national support. But whatever the mission, it must be compelling and the goals must be attractive.[23]

Throughout that spring, he presented numbers to Congress showing that NASA programs had created more wealth than even the military programs of World War II.[24] The history of technology repeatedly bears witness to the enormous value realized but not anticipated at the time of an undertaking, he said. People wanted to know from the leader of NASA why America should go into space. The answer was usually not well received. He emphasized that one must first realize that reinvesting the technology created by the space program led to jobs outside the space program. His position was that the new opportunities not only addressed poverty in the near term; they dealt with it for the long term and at the roots.

When the media challenged him on these arguments on the eve of Apollo 8, he was prepared. If difficult endeavors had to stop so that social ills might first be cured, he said, then everything would come to a grinding halt:

> Columbus would never have set sail, the Pilgrims would never have left home, and history, the arts and science would all be very different. . . . We simply could not hold up all progress and wait for all of society's problems to first be vanquished. . . . We must worry about how we create new wealth as much as we do about how we better distribute today's wealth. . . . It is a very sloppy intellectual argument to say that because there are poor people you cannot

> afford art or space or science. But don't argue we have the poor because we are spending money on space or weapons. All we argue is, let's take a very tiny fraction of our national wealth and invest it in pioneering the space frontier."[25]

In the last week of January 1969, George Mueller told him that the Apollo 9 lunar module was ready. Paine wanted to be sure and had asked Mueller to come across the street to his office. He was still uneasy about the flight-worthiness of the moon lander. The LM was an ungainly, fragile-looking vehicle designed to operate only in the vacuum of space. Since it did not have to fly within the confines of an atmosphere, the LM *was* fragile.

Originally called the Lunar Excursion Module by its designers, its development got off to a late start when it was the last of the prime contracts awarded on Project Apollo. In September 1962, the Grumman Aircraft Engineering Company of Bethpage, New York, had won the hotly contested contract over eight competitors. Their job was to design and build a spacecraft that could land two astronauts on the moon, stay on the surface for forty-eight to seventy-two hours, and then return them to lunar orbit.

Numerous delays had beset its development. Most had to do with the weight of the vehicle and combustion problems with the ascent and descent stage rocket engines. The first unmanned test flight, which had been scheduled for low-Earth orbit (LEO) in April 1967, did not take place until January 1968, and when it did, the flight revealed additional reliability issues with the engines. NASA had no choice but to postpone a second test flight that was scheduled for later in the year to give Grumman more time. Finally, in February 1969, the LM was ready for a manned test flight.

Once the launch date was finalized for the first week in March, Paine extended a personal invitation to Nixon to attend the launch. The president declined, but sent Vice President Agnew to the Kennedy Space Center. From the Launch Control Center three miles away, they watched Apollo 9 leave Launch Complex 39A for a ten-day mission to LEO. It would test out the complete Apollo hardware for the first time. In the years since, the flight has been largely forgotten (mostly because it never left Earth orbit), but it was as crucial as any of the others. It convinced Paine that the LM was, in fact, spaceworthy. The complex orbital approach, rendezvous, proximity operations, and docking as tested on the flight would all have to work in lunar orbit starting with the Apollo 10 mission.[26]

Upon splashdown, Seamans (now Nixon's new secretary of the Air Force) congratulated and at the same time challenged Paine: "I am grateful to know that the successful outcome of Apollo 9 is being confirmed. It is a truly historic prologue to the even more advanced accomplishments which we expect under your supervision."[27] With the six months saved from Apollo 8, the program was now back on track and actually slightly ahead of schedule for an attempt at a landing by the summer of 1969. Paine was soaring. His reputation with the new president and within NASA could not have been better.

While the mission was successful and achieved all of its major objectives, he still found himself defending his agency, this time in an unexpected public relations matter that had nothing to do with the outcome of the mission. Since Apollo 9 involved two spacecraft, call signs were needed for the first time. Astronauts Jim McDivitt, Dave Scott, and Rusty Schweickart had named their CSM and LM *Gumdrop* and *Spider*. It was an irreverent reference to the physical appearance of the two vehicles. The agency's Public Affairs Office was inundated with calls demanding to know who had come up with the inane names *Gumdrop* and *Spider*. This would happen again on Apollo 10 for the even more cavalier cartoon names *Charlie Brown* and *Snoopy* the crew gave their spacecraft.

Paine's position was that the criticism was totally unwarranted and rather frivolous, and that the media was trying its hardest to invent news to criticize about the country's very serious effort to land a man on the moon. Since he was asked about it, he had to respond. Like ships, he said, spacecraft also should have names, but the choice of the names was not what he considered most important in the larger project of going to the moon. "So often, people outside of the establishment feel that there is always tortured logic and debate and haranguing and bickering on these things," he said. History would come to embrace *Columbia* and *Eagle* on the historic flight of Apollo 11. "I was agreeable to having them name it anything they wanted to," explained Paine later that year. "If Neil [Armstrong] wanted to name it the *Jan Armstrong* [after his wife at the time], it would have been alright with me." He remembered, however, his regret that history would record no names for the Apollo 7 and 8 command modules, since there was only a single spacecraft on those missions and thus no need for separate call signs.[28]

Before Apollo 9 flew, Paine had to consider the possibility of skipping directly to a landing attempt on the next flight should the mission turn out to be a success. Mueller had brought up the possibility. That next flight, Apollo 10, was slated to be the final dress rehearsal for landing. It would test out every phase of a moon landing mission but the final powered descent from 47,000 feet to the lunar surface. With seven months remaining before the end of 1969, however, Paine told Mueller that there was simply no reason to skip directly to a hasty attempt at a landing. Besides, the Apollo 10 lunar module was too heavy and not equipped for an actual landing. (While all of the lunar modules looked similar, each vehicle was actually unique in what it could and could not do.)

He did, however, approve a major change in the flight plan for Apollo 10. The original plan had the LM simply separate from the CSM, maneuver away, rendezvous, and return for docking. In essence, it would have been a repeat of Apollo 9 but carried out in lunar orbit. Following Apollo 8, chief mission planner Howard W. (Bill) Tindall Jr. and his Mission Planning and Analysis team in Houston began lobbying for a more aggressive flight plan that would fly the LM down to 47,000 feet, the point just before it commenced its final approach to the surface. The descent stage engine had experienced "chugging" (intermittent combustion instability) on ground tests, signs of which had again appeared on Apollo 9. Tindall and the others all wanted to see if the problem would show up on a realistic trajectory burn. A revised, more daring flight profile would also check out the abort guidance system to simulate an abort during landing.

Phillips, Mueller, and Low presented Houston's plan to Paine. He deemed the additional risk manageable. The revised flight plan, albeit more aggressive, was still within the original scope of the F mission. Knowing NASA could have two more chances to attempt a landing before the end of the calendar year, Paine signed off on a May launch for Apollo 10. But he still had some reservations.[29]

The daunting fact that two spacecraft would be conducting maneuvers in lunar orbit three days away from any possible emergency return concerned him. Before Apollo 9, NASA had discussed flying the lunar module LM-4 on Apollo 10 without a crew to make a robotic landing on the surface. It would be a complete mission without putting a crew at risk. He favored the idea and pushed for it from NASA Headquarters after having discussed it at length with Low. But even after Apollo 9 flew with no major problems, he was still speaking with Mueller of flying an unmanned LM down to the surface. By now, however, his was the only voice lobbying for this position.

His caution was not unjustified. In the previous twelve months, two lunar landing training vehicles had crashed at Ellington Field in Houston during practice landings.[30] He told Mueller that his worst fear was that it might happen to a real lunar module on a real mission—that is, a crash onto the surface while trying to land. In March of that year, Low flew to Washington, met with Paine, and explained to him at length that the causes of the accidents were unique to the training vehicle itself and in no way connected with the actual spacecraft. But Paine still wanted the robotic option kept open. "If 10 goes wrong, we could still do 11 unmanned . . . If 10 is highly successful, we will land 11," he said five days before Apollo 10 left Launch Complex 39B. "I will be satisfied if we land by December."[31]

By the time Tom Stafford, John Young, and Gene Cernan splashed down in the Pacific on May 26, 1969, he needed no more convincing. He was squarely in the same camp with Mueller, Low, and the others. All of NASA was now eyeing July, not November, for the date when the United States would attempt to land the first human beings on the moon.

8

A GREAT SENSE OF TRIUMPH

I don't remember the occasion, but he told us that if we were unable to land on the first attempt, we could fly the following flight and try again.
—Neil Armstrong, recalling to the author his last conversation with Tom Paine before the flight

"He is wrong!"

Paine was emphatic. Ted Kennedy had clearly crossed the line this time. With three astronauts 230,000 miles away and about to enter into lunar orbit, NASA's biggest critic in Congress was publicly questioning why the United States was even going to the moon at all. The senator from Massachusetts had decided to use a memorial banquet for the late rocket pioneer Robert Goddard in Worcester as the backdrop for his harshest criticism yet of the country's manned space program.

"Frankly, we are surprised and disappointed that at a ceremony honoring Dr. Goddard, Senator Kennedy would present such a dispiriting vision of this nation's vigor and destiny in space," Paine replied, using a national press conference to address the senator. He was at Mission Control in Houston for Apollo 10, and the timing of the remark in the middle of the flight did not sit well with him at all. As he ended the press conference, he turned around and added, "We won't be including this item in the daily news reports we send up to the astronauts."[1]

The irony of Senator Edward M. (Ted) Kennedy's opposition to the US space program always puzzled him. The senator's older brother, President Kennedy, was the one who had made the bold gamble to send a man to the moon. As a national tribute, the spaceport on Cape Canaveral, Florida, from which astronauts left the planet was named in his honor. Now, as the country was about to make history, he was finding just how disruptive the younger Kennedy was. Senator Kennedy would oppose Paine his whole career.

Shortly after Apollo 10 splashed down and before Apollo 11 launched, he went to Kennedy. He proposed a truce, and offered an act of some sort that might link the American moon landing with the Kennedy name. The monumental significance of the event that was about to take place in just over a month would be unique in the annals of human history. Paine suggested that a simple gesture, perhaps leaving a personal memento of Jack Kennedy (a PT-109 tie clip) on the moon, might be fitting. He had discussed this with Julian Scheer, and the NASA Public Affairs Office thought it would indeed be a very appropriate, non-political unifying gesture. Another idea that they had come up with was to leave a Kennedy half-dollar on the surface. This, however, did not seem as fitting to him, since a piece of money would have hardly been the appropriate artifact to leave on the moon.

Kennedy, however, stonewalled Paine in the meeting and gave him no encouragement at all. Paine realized then that the Kennedys had no real interest in identifying JFK with America's moon program. If what he heard from Ted Kennedy was right, the family was, in fact, treating it as an aberration in the late president's legacy.[2]

His conclusion was later reinforced a second time. He had Scheer contact the Kennedys a year and a half later, in August of 1970. He wanted to see if they might consider accepting a moon rock from NASA to be placed at the JFK grave site in Arlington National Cemetery. Paine had no personal allegiance to the Kennedys, but wanted to try to give what he believed to be a rightful tribute to JFK. The late president was the one who had moved the nation forward in a time when many doubted America's resolve to do something historic.

As before, Ted Kennedy rejected the offer, saying the family would be very much opposed to such a move. The Kennedys had turned down requests before to emplace mementos at the grave site. They were already distressed at the large number of tourists disturbing the serenity of the site in the years since the assassination. The senator expressed an interest, however, that NASA might perhaps bestow a moon rock on the John F. Kennedy Library and Museum—with the caveat that it be stored indefinitely in a Boston office until the presidential library opened sometime in the future. This doused Paine's enthusiasm completely; he had no desire at all to see a piece of the moon tucked away in an office safe somewhere in downtown Boston. He gave up talking to Kennedy after that. The reluctance of the family to associate the space program with President Kennedy disappointed him greatly.[3]

It was now the Sunday afternoon of the long Fourth of July weekend. Paine and Associate Deputy Administrator Willis Shapley were talking in his Washington office. Their attention was focused on a drawing that Bob Gilruth had just sent over by Telecopier (fax) from Houston. The launch of Apollo 11 was now just ten days away. This was the last opportunity to make any changes to the brushed stainless steel commemorative plaque that the astronauts would leave behind after landing on the Sea of Tranquility.[4] The plaque would mark the spot where mankind first touched down on the surface of another celestial body. It was one of the historical accoutrements of the flight that was not without controversy.

An ad hoc Committee on Symbolic Activities for the First Lunar Landing, which Paine had appointed back in February, had recommended that Neil Armstrong and Buzz Aldrin do two notable things as part of their historic moonwalk activities. The first was to plant the US flag;[5] the second was to unveil the plaque. He had polled the directors of the NASA field centers from across the country and asked them whether the flag of the United Nations should be planted, and if so, whether it should be the only flag planted or whether it should be planted alongside the Stars and Stripes. The center directors told him unanimously: plant only the US flag. As for the UN flag, "You do that and you're asking to be crucified," Hans Mark, the former director of the Ames Research Center, put it. "Don't do it," he recalled. "The American taxpayers paid for it, and dammit, if you put a UN flag next to that, you're going to have half a nation angry at you. I don't know what my colleagues did, but I sure as hell was very hard over on that," he said.[6]

But the controversy over the commemorative plaque was strictly partisan Republican versus Democrat politics. When it was first proposed, the plaque (which was attached to the front landing strut of the LM descent stage that was left behind on the moon) was to bear a print of the US flag. It was then changed to show the eastern and western hemispheres of Earth, along with the inscription "We come in peace for all mankind." The signatures of astronauts Armstrong, Michael Collins, and Aldrin had also been added by Houston. Paine and Shapley looked at the new design. Shapley liked it, but thought it was still missing something. He could not quite pinpoint what it was, however. They played with a few pencil sketches but could not improve on what Houston had designed. It was getting late, and so they sent it over to the White House.[7]

The White House came back with two additional changes the very next morning. (It was about the fastest response that he had ever received from the White House, Paine later recalled.) First, they wanted to alter the message "We come in peace for all mankind" to "We came in peace for all mankind" to tone the message down from sounding heavy and perhaps a bit hostile, to one more of historicalness. The second change was that Richard Nixon wanted his signature on it.

President Nixon received a lot of criticism for wanting to put his name on the plaque. Many derided him for signing it, but Paine was not one of them. He instead defended the action of the president, who he believed "had every right to sign that in the name of all the people . . . . We were not getting the signature of Dick Nixon from Whittier, California. We were getting the signature of the President of the United States, and [by] the President of the United States signing that, we were getting the Chief Executive Officer of the entire 200 million Americans who had done this thing to sign on that. . . . I thought the press completely missed the point and were using this as really an undercover attack on the President."[8] He viewed the plaque not as a political symbol but as a historical token. To him, the president was the one person that history could point to as the head of state of the country in the epoch when human beings first reached the moon. Whether that was fortuitous happenstance or not was not the point. Angering the Democrats even more was that Richard Nixon's signature, in fact, turned out to be the only presidential signature left on the moon. Only the Apollo 11 and 17 plaques (the first landing in 1969 and the last landing in 1972) bore the president's signature.

Nine days later, Paine found himself standing in the middle of a field surrounded by a group of reporters and cameras. He was engaged in a lively discussion with the Reverend Ralph Abernathy, but their conversation was no means private. The dark clouds of a nearby thunderstorm added to the drama of the confrontation. The civil rights showdown on the eve of the launch of Apollo 11 was one that he knew was coming but could not avoid. The American civil rights leader from Alabama had become the head of the Southern Christian Leadership Conference's Poor People's Campaign after the Reverend Martin Luther King Jr. had died in his arms from an assassin's bullet on April 4, 1968. Scheer had told Paine to be ready for Abernathy. Just a month before, Abernathy had gone to Washington and was jailed for failing to leave a march. Now he, along with Hosea Williams and members of his mule train hunger marchers, had descended en masse on the Kennedy Space Center. They were there to protest the fact that America would go to the moon while so many were still living in poverty.

With the nation watching, Paine and Scheer met Abernathy just outside the main gate of the space center. The timing of the protest bothered Paine. His position on civil rights was that the complex social problems facing the nation, problems like the ghettos, racial discrimination, crime, and poverty, were not things that could be solved by not launching the Apollo 11 flight to the moon. The two issues were mutually exclusive, he often said. If all of America's ills could be solved by not pushing the button to launch the rocket, as he put it, he "would not push it."

Since Abernathy's people had by then camped outside the space center, Paine sent a bus the next morning, served them breakfast, and brought the group to the VIP viewing stands. There, Abernathy, Williams, and the others watched the moon launch with dignitaries and guests from around the globe. He had no quarrel with Abernathy and never forgot what the reverend told him that day. Although Abernathy conceded that he was using the drama of Apollo 11 to focus national attention on what he felt were major failings of equality in social injustice, nobody in this country could be prouder of the astronauts' achievements than he was. That evening, he held a prayer meeting and the marchers prayed for the safety of Armstrong, Collins, and Aldrin.[9]

The attention of the world was focused on NASA on July 16, 1969. That was why Abernathy was there. Paine was sympathetic to their cause, but his chief concern was to contain the situation:

> The thing that really bothered me when the Abernathy thing came up down at The Cape was that if we allowed [his] movement to contrast the situation of hunger and poverty in the United States with the Moon launch the next day, that we would be doing a very obvious disservice, I thought, to the NASA program. If a great deal of publicity were to be given to this, and particularly if the lunar program were not successful and it could be contrasted with the fact that here we have spent all this money and we perhaps were not able to get to the Moon, it may be a fiasco or even a tragedy. Contrasting it then with Abernathy's point that the money should have been spent somewhere else, it certainly could do grave disservice to NASA and even to the nation that invested $20 billion in this venture. So I was very concerned that we handle the . . . situation in the coolest and most correct possible fashion to minimize any reflections, any ability that [he] might have to seize the stage and hold it to the detriment of the lunar program. Even though I had a good deal of sympathy with Abernathy's objectives, it was very important that we not allow it to do harm to the NASA program . . . and put it in the best possible light of the space agency.[10]

He had to answer for the nation's space agency. The question of whether the billions spent going to the moon would have been better spent on poverty programs was one that he tried to answer with conviction. He wanted to be as pragmatic as he could. There was no easy answer, as the debate appealed to people's emotion. His encounter with Abernathy showed this. The dilemma was not unique to the space agency, however. It also applied to other parts of the federal government. The Department of Housing and Urban Development

(HUD) and the Department of Health, Education and Welfare (HEW), for example, had to be held accountable to the American people as well. To this end, he pointed out every chance he had that the two much larger departments, whose job it was to directly address and tackle the country's social issues, were never subject to the same level of scrutiny as the space agency.

The $20 billion (over $100 billion in 2017 dollars as adjusted for inflation) that NASA spent on Project Apollo equaled a mere 4 percent of the half a trillion dollars that HUD and HEW spent on social programs during the 1960s. "If we had foregone US advances in space and science technology, if we had allowed the new ocean of space to become a Russian lake, our cities and social problems would be just where they are today. But what a miserably defeatist view of America is implied by the view that our mighty nation can only do one thing at a time."[11]

> We must do many things at once. It is just as important to carry the garbage and sewage out of a modern city as it is to have police and fire protection, to have beautiful buildings and fountains, to educate the children, to support art museums and baseball teams. . . . We have got to do all of these things and many others simultaneously. . . . The space program in the United States is less than a-half-of-one-percent of our gross national product. . . . Unfortunately, the public does not realize what a very modest percentage of the resources of America—the wealthiest nation in history—is devoted to our space program. To argue that by reallocating this modest effort we could somehow create utopia is obviously fatuous, as is the emotional reasoning that balances NASA against poverty. Stopping the space program would only stop U.S. progress in space, that's all.[12]

By the beginning of 1969, the progression of flights had put Apollo 11 in line to make the first landing attempt. At NASA Headquarters, Paine and Mueller generally stayed out of the crew selection process, deferring to Deke Slayton and the Flight Crew Operations Directorate in Houston to put forward the best crew for each mission. This included

Apollo 11. By the end of 1968, it was becoming clear to everyone in NASA that if the schedule held, Apollo 10, 11, or 12 would attempt the first landing, with 11 being the most likely in the group. Since crew training requirements had to be identified over a year before a flight, Slayton had assigned NASA's most experienced flight crews to begin preparations so that any one of them could be ready when the time came.[13]

Timing had put Paine in charge of NASA. At the time, he knew Neil Armstrong only as well as he did other members of the Apollo flight crews—in their capacities as NASA boss and NASA astronaut. Only after the flight did Paine get to know him much better on a professional level. Paine did not make Armstrong the first man to set foot on the moon, but he did sign off on the decision.

Initially, the topic of who would be the first man on the moon was not directly addressed by NASA managers. It began to take on a life of its own, however, as the national press made it into a worldwide sensation right after the Apollo 11 prime crew was named that January. There was wide public speculation that Armstrong, as the mission commander, would be the first rather than Buzz Aldrin, who was the lunar module pilot and the third-ranking member of the crew behind Armstrong and Command Module Pilot Michael Collins. But this would have broken with the pattern of the copilot (Aldrin, in this case) leaving the spacecraft on space walks. This had been the practice on five previous Gemini flights and on Apollo 9. On those occasions, however, the commander did not leave the spacecraft, but instead stayed inside to monitor activities. Since both astronauts would now be leaving the lander to explore the lunar surface, this protocol no longer applied.

Armstrong's version of the decision that made him the first person out of the spacecraft never changed in the years after the flight. When asked about it for this biography, he recounted again what he believed took place: that technical factors drove the decision. Hours of simulations in Houston that spring had experimented with how he and Aldrin could best don their bulky portable life support systems, open the hatch, and position themselves inside the cabin to back down the ladder of the lunar module. Engineers evaluated the advantages and disadvantages of the various ways that they could exit (and, even more importantly, reenter) the spacecraft. The exercises showed that the best way to egress the very limited confines of the lander was to have the commander go first. This affected not only who went first but the tasks that each astronaut performed while on the surface. George C. Franklin, subsystem manager for the LM crew station at the Manned Spacecraft Center, had detailed these factors in a technical report to Slayton.[14]

When Slayton offered him the command of Apollo 11, Armstrong had first picked Collins and later added Aldrin—in place of James Lovell, whom Armstrong totally agreed should have the opportunity for his own command, something he later got on Apollo 13—from, in his words, "the relatively small pool" of astronauts who were qualified and available for mission assignment at the time. Of the twenty-four astronauts fitting this category in the Astronauts Office at the time, only two were veterans of at least one Gemini flight: Richard Gordon, who flew on Gemini 11 with Charles (Pete) Conrad, and Aldrin, who flew on Gemini 12. Conrad, whom Slayton had already picked to command Apollo 12, now wanted Gordon as his command module pilot on Apollo 12. This left Aldrin to round out Armstrong's crew on Apollo 11.[15]

The question of who should be first on the moon was historically important only on Apollo 11. In 2001, Chris Kraft revealed that he, George Low, Deke Slayton, and Bob Gilruth had discussed the issue and made a decision favoring Armstrong over Aldrin. This was recounted in James Hansen's authorized biography of Armstrong, *First Man*. The technical perspective, while quite valid and important, ended up justifying the shrouded management decision. With Franklin's report in hand, Gilruth conferred with Sam Phillips and George Mueller in Washington. He argued that technical considerations, followed by command prerogatives, affirmed that the commander should be the first to climb down the ladder. Phillips and Mueller had no problems with that. Mueller then notified Paine of his recommendation in a routine Monday-morning memorandum to the administrator. Three and a half months before launch, on April 7, Paine signed the internal memorandum and made it official. Neil Armstrong would be the first man to step on the surface.[16]

July 20, 1969. Civilization realized its age-old dream as Armstrong and Aldrin piloted *Eagle* to a soft landing on the Sea of Tranquility near the equator of the moon.[17] Six hours later, the world saw the drama continue, as fuzzy black-and-white television images recorded the first human steps on the pristine surface of the moon. Like the rest of the audience watching the EVA (moonwalk) on television, Paine did not know what the first spoken words of a human being on the moon were going to be. The subject never came up—not until Armstrong's unflappable "one giant leap for mankind," as the quietly confident thirty-eight-year-old engineer and test pilot from Ohio gently placed his left boot on the surface and into human history.[18]

It capped off a decade's worth of work by three hundred thousand people in the US workforce. Human beings' first excursion on another world was brief, just two and a half hours from start to finish. Armstrong and Aldrin gathered rock samples, carried out a few simple experiments, set up a television camera, planted the American flag, unveiled the plaque, and took an unscheduled telephone call from President Nixon.

Four days earlier, as the mighty Saturn V ripped through the azure Florida skies, not unlike a majestic monument leaving Earth, Tom Paine had momentarily let his veil down to reveal "a great sense of triumph."[19] Now, as he watched mankind's first steps on the moon from the Mission Control Viewing Room in Houston, he thought the world's view of America changed. As with Apollo 8 seven months before, America's standing in the world had been reaffirmed. It was, in a way, the culmination of the United States reminding the world that it was still the great nation that it believed it was before *Sputnik*. Whether or not it was worth the $20 billion price tag was something that had to await the judgment of history. But in his mind, there was no question that Apollo was well worth it.[20]

He had said shortly before launch that the sight of two Americans walking on the moon would do wonders for the nation's self-esteem. It was the conclusion of a decade when many Americans openly questioned and demonstrated against their country. The feat had forced America to rethink its position among the nations. The landing pushed ahead almost every facet of science and engineering. Nearly everything that modern technology was making possible had to be extended a step further in order for the United States to land on the moon. And most importantly, America chose to pursue that challenge.[21]

The flight of Apollo 11 was still more remarkable to him because the country chose to be transparent in doing it. It broadcast the mission live to a global audience as it happened, minute by minute. Paine attributed a great part of its success to people who persevered through contentious political differences to accomplish a national goal. In the weeks after Apollo 11 returned safely to Earth, he gave credit publicly to many people inside and outside the space agency. It had all started with Senator Lyndon Johnson and Representative John W. McCormack drafting the Space Act. Dwight Eisenhower then established NASA. John Kennedy challenged a nation. Congress did its part also; Representative George P. Miller and Senator Clinton P. Anderson, in particular, gave unwavering support from Capitol Hill for the space agency to stave off what was otherwise a crippling budget retrenchment. Richard Nixon continued with the program when he could have canceled it. But Paine reserved his highest praise of all for Jim Webb, who had led the agency through what was arguably the most trying time of its existence in the months following the Apollo 1 fire.[22]

Project Apollo galvanized a nation, if only briefly. Even though Paine did not subscribe to the popular interpretation of the Space Race, fundamental differences between the US system of government and that of the Soviet Union were obvious, as the way the two countries approached their lunar programs had so vividly shown. Both superpowers possessed the people and the resources that were needed to make progress in space; both had assembled the teams of people and the infrastructures needed to succeed. The real question, however, turned out to be one of national daring, national will, and national resolve. Here, he thought the disparity was clear. The Soviet Union emphasized progress in space as an external demonstration of the superiority of the communist state; being first on the moon was something they very much wanted for that reason. That was why right after the American moon landings, the Soviets forged ahead with their (unsuccessful) Mars program.

Paine understood that the Soviets might very well regain the advantage in space after Apollo. But if they did, it would not be due to any technological breakthrough. Rather, his concern was that the United States might feel that, having landed on the moon, it could now relax and rest on its laurels. There was also the national security aspect. "It would now be very unwise of us," he said, "to allow any other nation to think that they have attained great superiority over us in an area as important as space. It would be a destabilizing force in the world today to have such a situation exist tempting others to rash action." In 1969, success in space meant America and the West were more secure here on Earth.[23]

The sky was still dim just before daybreak on July 24, 1969. Nixon was basking in the limelight. He was fraternizing with the sailors, waving to everyone who saw him, and generally enjoying being on the bridge of the USS *Hornet* as it waited for command module *Columbia* to splash down as it returned from the moon. Apollo 11 was almost home. Paine, standing beside Nixon, was nervously rehearsing in his head the final sequence of events during reentry blackout. A lot could still go wrong in the next ten minutes. He remembered conversations he had had with engineers when he was still the deputy administrator. What concerned him most was the possibility that the parachutes might not deploy.

> My own concern was very much the fact that we had carried this tremendously risky, hazardous, unprecedented mission all the way through to reentry, and I was just hoping to God that the remainder was going to be successful. I was thinking very fondly of Eberhard Rees and the work that he had [done] on the flowerpot assembly [parachute cluster] at the top of the Command Module, and the work that he had done on the parachutes and the fact that [he had] assured me that this problem was really solved and that we could bring that thing down with one drogue and two parachutes, if necessary, and that everything possible had been done. I was going back over in my mind asking myself was there any question that we had left unanswered.[24]

He looked up into the dawn, searching intently for any trace of *Columbia* as it made its fiery reentry toward the Pacific some 240 miles south of Johnston Atoll. Most of the sky was obscured by clouds, but there was a very small opening almost directly over the carrier. His eyes dashed back and forth between that opening and the second hand on his wristwatch, anticipating the sight of the capsule at the appointed time and hoping to be the first to catch a glimpse of *Columbia*. An orange glow finally appeared. It quickly turned into a bright burning streak not unlike an unusually bright shooting star. He grabbed Nixon by the arm and said, "There it is. There it is. Here they come, Mr. President!"[25] Moments later, it disappeared behind the clouds. But by then, he was pretty certain that they had done it, as the timing of the reentry was so precise. "That was my last lingering doubt, and I was very confident then that we should shortly pick up the chutes. I began to relax, even though our usual dictum is [to] not relax until the wheels of the helicopter are touching down on the deck of the carrier." Fifty-five minutes later, *Helo 66* was back on the deck of the *Hornet* with Armstrong, Collins, and Aldrin. He could relax now.

Richard Nixon had already planned on an eleven-day world tour with First Lady Pat Nixon. But only after *Eagle* had successfully landed on the moon did he tell Paine that he was going to kick it off by flying to the middle of the Pacific to see the splashdown. Paine thought that it was a "rather bold move" on the part of the president. There was a definite risk in the decision, and Paine wondered whether Nixon knew the odds as well as he. "That was all very well and good the way it turned out," he later said, "but I was scared to death that we would have a fiasco or even a tragedy, and that this entire presidential goodwill mission would be scrubbed and that the initial impact to the space program would be something like the U-2 affair was on Eisenhower's great trip to bring peace to the world in Paris. . . . President Nixon certainly put us on the line."[26]

The effusive global response to the flight of Apollo 11 poured in to his office. He also saw firsthand on his travels overseas the affection for the American space program. People widely and often referred to it as "their" program. This was especially true in places like Australia and Spain, where

NASA had network tracking stations. Western countries that had anything to do with Project Apollo adopted the US space program as their own. He found almost everywhere he went that the American space program was regarded as planet Earth's space program, in which all peoples participated through the technology of live television. NASA became a symbol of success, and Apollo 11 became a symbol of triumph for all mankind. But when Paine spoke, he always reminded people not to forget that the feat was carried out by the American workforce and funded by the American taxpayer. While there was global participation, it was America that had made that triumph possible.[27]

Speaking to the National Press Club after the mission, Paine attributed the overwhelming global reaction to NASA's unique founding charter. By choosing to be open and transparent, the country had conferred a tremendous and far-reaching effect on the world opinion of America. To this end, he contrasted the ideological differences between the United States and the Soviet Union. Yuri Gagarin, the first human being in space, was put into orbit by the communist state in 1961—by all accounts an impressive feat, rightfully credited to them. But the closed, secretive nature of the accomplishment prevented the world from joining in the triumph. There was no way for other countries or even the Russian people themselves to share in the human drama, as Moscow tightly controlled the dissemination of any such news. The history-making Apollo missions provided a podium for the United States. By being open, the nation had earned that respect.[28]

Here lay a fundamental difference between what Paine and his predecessor Jim Webb believed drove the American space program. Webb wanted to stay ahead of the Soviet Union, beat them to the moon, and win the Space Race. By all accounts, this was a successful and necessary formula.[29] While Paine acknowledged that there was a competition with the Soviets, he downplayed it, and never pointed to it as the reason that should ever drive the US space program. Instead, he advocated and would continue to advocate for a vigorous space program based not on reaction to what others might do but based on the intrinsic worth of the benefits of human exploration to society.[30] It was a noble argument that was, however, not pragmatic enough to provide a rationale for funding a federal government agency.

Paine compared mankind's venture into space in the twentieth century to the Europeans leaving the Atlantic coast in the fifteenth century. Then, kingdoms and kings had to ask themselves how much they should invest in this new and unknown frontier. For example, the Portuguese voyages of discovery had far-reaching effects beyond sailing the seas. They provided a focus for the best that Europe offered, from cartographers to shipwrights, coopers, and gunsmiths. Lisbon became the richest city in Europe. The commitment to exploration and maritime technology later spurred the mighty Spanish, British, French, and Dutch empires. Their voyages changed history because they were ambitious. Extraordinary results came from setting high goals.[31] History has repeatedly shown that exploration of the unknown always produces unanticipated, and often immense, long-range benefits. The quest for pearls, spices, and precious metals motivated Columbus. The Spaniards came to the New World to extract wealth. But in the long run, the value of the gold taken from the Americas proved infinitesimal compared to the new society that was formed. The voyages changed history. The outcome was extraordinary because goals were set extremely high. The difficult challenges caused the best and the most enthusiastic to achieve beyond the normal. This was what NASA had provided the country by going to the moon: "a national value without a price tag." For Paine, this was the true value of the space program.[32]

In the months following the flight, open dates on his calendar were few and far between. The White House had arranged for a monumental, one-day national celebratory tour for the crew to take place as soon as they had cleared quarantine—a strictly precautionary measure, in case some unknown moon bacteria came back with them. On August 13, Armstrong, Collins, and Aldrin boarded *Air Force Two* with their families and. along with Tom and Barbara, flew from Houston to New York, Chicago, and Los Angeles. Julian Scheer had objected to the ridiculous pace of the schedule, but the White House won.

In New York, the epic scale of the ticker-tape parade through the "Canyon of Heroes" in Lower Manhattan was one that the city had not been seen since V-J Day at the end of World War II. Riding in the convertible under the rain of confetti and watching the three astronauts waving to the crowd, Paine smiled through it all. They spoke at City Hall, where Mayor John Lindsay presented them with keys to the city. The motorcade continued to the United Nations, where they were greeted in a commemorative ceremony hosted by Secretary General U Thant.[33]

Arriving later that morning at the Chicago Civic Center, Paine introduced the astronauts to the crowd and then stepped back and applauded with a crowd estimated at 3.5 million that had come from all over the Midwest. Illinois Governor Richard B. Ogilvie proclaimed Armstrong, Collins, and Aldrin the "First Citizens of the New Epoch."[34] After they addressed fifteen thousand young people in Grant Park, it was off to Los Angeles. Richard and Pat Nixon, just back from their eight-country goodwill tour, were waiting at the Century Plaza Hotel. The evening saw the largest state dinner ever hosted by a sitting president. (It was also a Secret Service nightmare, recalled Paine, who along with Wernher von Braun, lost his assigned seat to others in all the confusion.) He presented the NASA Distinguished Service Medal to several people; Agnew presented the Medal of Freedom to Armstrong, Collins, and Aldrin; and Ronald Reagan, then governor of California, told the astronauts that they were not getting any special medals from him as door prizes for having traveled the greatest distance for dinner. The Reverend Billy Graham closed the evening with an adulating and moving benediction.[35]

The hectic pace continued. On September 7, Paine flew to New York City, where the next day he addressed the United Nations Committee on the Peaceful Uses of Outer Space. It was a large moment for the space agency, the United States, and for him. The US space program had made history. It was a harbinger of things to come, he prophesied in a speech that day heard in eight languages inside the General Assembly Hall. One day in the not too distant future, the vast frontier of outer space would provide another kind of hope for mankind, he said with great passion and optimism:

> Apollo 11, a most dramatic extension of man's capabilities in space, was an achievement by and for peoples everywhere. This event has implications for mankind far richer and more meaningful than a landing on the Moon in the narrowest technical

sense. If men properly develop and exploit these advanced capabilities, they can surely be directed to a great expansion of those practical benefits which we have only just begun to reap in space in the fields of communications, weather prediction, navigation, Earth resources, and other fields. Man will be able, in time, to extend his domain beyond the confines of his home planet Earth. From our small 8,000 mile diameter planet, we have set forth in this first step upward and outward into the 8,000 million mile solar system around us.

When I say that the success of Apollo 11 is a step forward of all mankind, I do not use these words without thought. The variety and extent of foreign contributions to the Apollo 11 flight are real and they are impressive and they are appreciated by all Americans . . . . If man's reach should exceed his grasp, the fact that we have been able, in the Apollo program, to grasp the Moon shows that man has perhaps not been reaching far enough. We can dare and we can win far more for man than we have ever thought possible. And, we should not only in science and technology but in all the affairs of men.

It is very proper that men everywhere around the world are asking us if man can indeed go to the Moon, why can't we do a far better job here on our planet Earth in ordering the affairs of man. This is a question which is indeed appropriate and a question which those of us concerned with space programs should welcome.

There is much to be learned in space and it is relevant to our total environmental knowledge here on Earth. We are opening a whole new field, that of planet ecology. . . . We can, and we must, pursue this increased knowledge, and we must turn it increasingly to the benefit of man. To equip ourselves for this task, we should continue the work we have begun and should increase our capabilities still further, but above all, we should do it as much as possible together.

After the Apollo program, we see a very rigorous opportunity to press forward. We believe that the Apollo 11 astronauts have opened a trail that many men will follow. Their flight is a beginning, not an end. We stand at the start of a new era which will see space flight become as safe, as reliable, and as economical as aircraft flight through the atmosphere is today. We see lying ahead of us now the task of developing reusable spacecraft and permanent space stations in orbit that will greatly reduce the cost of space operations and will open space travel to men and women of all nations. The future space programs will consist of equipment that will be multipurpose, it will be used many times and will bring back in many areas far more information than we have been able to acquire in the first dozen years of space.

These future programs can and should be carried forward with far greater international participation than has yet been the case. That participation will be as rewarding to all nations who take part as it has been to those nations which have started down this trail. The character of the space effort in the name of all mankind will surely be more rewarding to every person on this planet and will well repay the energies and the resources required. Certainly, we in the United States will, as we have in the past, make increasing opportunities available to people of all nations who wish to join with us in the pressing forward of this great human endeavor.

The great exploration of history, carried out by many nations, [has] always opened up new vistas of the possible, and the sights of all men have been raised and their hearts inspired. The exploration of space is in that great tradition, and yet it extends by orders of magnitude the past explorations. Where before Apollo exploration was a challenge in itself, its successful beginnings now stand as a challenge for our children and for all future generations as we open up this limitless frontier. Certainly the greatest challenge of all is that the world, which is seen as one from space, should also be seen as one from the Earth itself.[36]

9

GOING GLOBAL

I feel it is only right that we should share both
the adventures and the benefits of space.
—Richard Nixon, addressing the United Nations General Assembly

It was one of those rare times when he sat down and talked at length with the president in the Oval Office, and the president was in an exceptionally good mood. Richard Nixon and Tom Paine could not have regarded the US space program more differently. Nixon was lukewarm, at best, toward the space program, but was happy as long as NASA played its part in his administration. Paine knew this all too well. To keep NASA on Nixon's radar after Apollo 11, he had to take a risk and be as bold as possible with the president. The White House was capitalizing fantastically on the triumph of the first moon landing. Nixon wasted no time. He sent Paine on a series of visits to nineteen countries in Western Europe and Asia beginning in October of that year. More of a guest than a dignitary, Paine took in the sights and sounds of Bonn, Paris, London, and Ottawa—capitals of the anchoring nations in what would soon be the European Space Agency. A highly publicized trip to Canberra and Tokyo followed in February of 1970.[1]

This goodwill tour was the best possible way that the president could use the returns from the space program to advance his international relations agenda overseas. Nixon's foreign-policy objectives in the early 1970s were deemed by Democrats and Republicans alike as broad and daring. His use of shuttle diplomacy as spearheaded by Henry Kissinger proved remarkably successful. It opened relations with the Soviet Union, the People's Republic of China, and in the volatile Middle East in a sweeping way. The president made it clear to Paine that, as the head of NASA, he needed to devote as much time as it took to building a strong relationship with the space agencies of the other countries. Above all, Nixon wanted him talking with the Russians.

Paine remarked once in a speech that having raised with Barbara four half-American, half-Australian children, he should be an expert on foreign relations. He was not a diplomat, nor did he know international law. Someone had to help him, so he called on Arnold W. Frutkin, NASA's assistant administrator for international affairs. Frutkin was the agency's senior negotiator for almost all international agreements from 1959 until he retired in 1979. In those twenty years, he defined a great deal of the agency's foreign-policy objectives. The guidelines still govern the operation of the International Space Station in the second decade of the twenty-first century. He and Paine got along fantastically.

Frutkin later remembered that Paine's travels drew intense scrutiny in the international community at the time. His speeches garnered a great deal of publicity overseas; long quotes and interviews were often reprinted in their entirety in newspapers and magazines.[2] He was now widely known in the space and science communities of Europe and Australia. VIP treatment that was usually reserved for prime ministers and state officials was not uncommon. Helping Paine and Frutkin with government contacts and advance planning as they visited each country was Richard J. H. Barnes, the agency's representative in Europe.

In the weeks after Apollo 11's landing, Paine spent a good portion of his time writing letters to the science ministers and officials of other countries inviting them to submit proposals for potential cooperation with the United States. The topics could be anything, he wrote. He would consider anything reasonable, from simple atmospheric research experiments to ideas for joint human spaceflight. The invitation surprised the recipients, in particular the Soviet Union. They had not expected the US to be so open. The impetus came directly from Richard Nixon. He told Paine that America needed to use the high visibility of the space program as a new global ambassador. If successful, the benefits of space sciences might be used to spearhead American-led humanitarian programs and bolster America's image on the world stage. The two moon landings (Apollo 12 had successfully followed Apollo 11 to the surface of the moon in November) had already lifted America's standing in the world. Nixon now wanted to maximize the returns.[3]

Paine was all for it. In the letters, he encouraged allies to think about the long-term benefits of working with the US. He was extremely optimistic about the space age. The opportunities for down-to-earth applications were broad and beyond the ability of the United States alone to exploit. NASA's accomplishments in space had already made an impact on more than just science and engineering. In the 1960s, it was actively changing how other nations viewed America by showing what a free society, galvanizing its will and focusing its resources on a common goal, could do. Moreover, it showed the peaceful intent of America in space. In going to the moon, the United States never threatened the security or vital interest of any other nation. If the US sought to garner praise from other countries, then Paine saw Apollo 8 as the one mission—even more so than Apollo 11—that first changed the way people viewed America because of her achievements in space.[4]

Responses to his invitation started to trickle in over the next few months. By the end of 1969, he had received quite a few responses. He had to be judicious in just how the agency treated a foreign proposal. He directed Frutkin to only pursue projects that were well defined and of clear mutual interest to both countries. Each nation had to fund its own responsibilities. In his words, both should gain, and neither should be forced into "giveaways."[5] For example, if France approached the US with a joint satellite proposal, the question would be, while the French might be interested in the science, was the US interested? Unlike most joint Department of Defense agreements, there was no exchange of money between the United States and an international partner. A foreigner sent to the States for training, for instance, had to be funded entirely by his or her own government.

Paine had learned a valuable lesson from Glenn T. Seaborg, the Nobel Prize laureate who was then chairman of the Atomic Energy Commission. The AEC (a forerunner to what is now the Department of Energy) was, at the time, sponsoring and paying for international training of all kinds of personnel from several US allies. Foreign scientists and engineers gladly came for training at the expense of the United States. They completed their training and went home with no further obligation. NASA, therefore, set up the ground rule that each side should do what it could at its own expense. If Canada wanted a cooperative satellite project, it would build and pay for the satellite, which the United States would then launch. With its global network of ground stations, the US could provide tracking and data acquisition services that Canada would then pay for. Since satellites generally cost quite a bit more than launchers, the Canadians would, in this case, bear the preponderance of the cost, thereby saving the agency money.[6]

In October 1969, he presented NASA's long-awaited plans for the space shuttle to the international community. It was a big moment for the space agency. Paine personally invited, by way of letters and phone calls, representatives from forty-three countries. They converged on Washington to discuss ideas. Most interested was the consortium of industrialized nations of West Germany, France, the United Kingdom, the Netherlands, Canada, Australia, Brazil, Sweden, and Italy. The European Space Conference, European Space Research Organization (ESRO), and European Launcher Development Organization (ELDO) all sent representatives.[7]

The following March, he convened a similar meeting. This time, it was for a study project on designs for a space station that would orbit Earth. The response was again overwhelming. Testifying before Congress on March 11, 1970, Paine read a letter he had recently received from Sir Hermann Bondi, the world-renowned cosmologist and director of ESRO. In it, Bondi told him that it was not the specifics of the program that most impressed him but that the US was serious at all about developing programs involving other countries. This came on the heels of Project Apollo, which had been an exclusively American undertaking. Paine hoped that by sharing information freely, openly, and at the inception of these programs, potential partners might choose to commit early funding toward these large undertakings.[8]

Three months later, in June, he and Frutkin flew to Paris. Before he left, he met with Henry Kissinger in Kissinger's West Wing office. Nixon's most trusted foreign-policy adviser gave him a clear set of instructions. It was very important to keep the European community focused on their prospects for cooperation he said. Smaller countries like Belgium and the Netherlands were starting to voice their wariness that their position might be dwarfed by countries like France and West Germany. There was also talk that some of the participating nations were about to enter into agreements with the US apart from the others.

Paine relayed a strong message to them: do not get bogged down in political differences. The Europeans needed to think about how the new opportunities, technologies, and missions of the American space program might work to strengthen the space alliance as a whole.[9] Following the Paris meeting, he and Frutkin boarded a train to Brussels and met with the Belgian science minister. This was followed by a stop at the Dutch Science Ministry. At The Hague, he signed a seminal agreement for the United States and the Netherlands to launch the joint Small Astronomy Satellites, a groundbreaking venture with the small European country.

On Thursday, September 18, 1969, he signed another landmark agreement. This time, it was with India's Department of Atomic Energy. The midday ceremony at NASA Headquarters received a surprising amount of international attention. It allowed the US to help provide the half a billion people of India with the nascent technology to do something that the rest of the western world had already been doing for years: watch live television.

The population of the world's second most populous country was spread throughout all regions of the subcontinent; rural television was still nonexistent. The State Department, however, had not initially endorsed the idea.[10] Having come through 150 years of British rule, India for years had refused attempts by the United States to establish a Voice of America radio station on its soil. Frutkin was able to convince Vikram A. Sarabhai, chairman of the Indian Space Research Organization, that the US was not interested in disseminating propaganda. Paine, in the meantime, spent many long hours meeting with officials from the US State Department. He made it clear that NASA satellites would be used to transmit only public education television signals. The project accelerated. As NASA made its satellites available to India, it was up to India to develop its own ground stations. They could also buy equipment from the United States. No Americans had to set foot on Indian soil. His message to the international community was clear: if they wanted to participate, they should seize the moment.[11]

As this was happening, President Nixon was banking on the possibility that multilateralism in space could further the détente with the Soviet Union. A temporary reprieve in East-West tensions had led to the signing of the Outer Space Treaty by the United States, the United Kingdom, and the Soviet Union in 1967. The Treaty on the Non-Proliferation of Nuclear Weapons had garnered over forty signatory states the following year. Timing was now of the essence.

He knew that Nixon was indifferent to space exploration per se, but the president understood its political value. He asked Paine to capitalize on the short window of opportunity created by the public's curiosity about space travel. He had seen that having astronauts in the news brought prestige to the country. It also made him look good. He once told a gathering of NASA officials that no president of the United States would ever want to stop sending men into space.[12]

This was at a time in the Cold War when any talk of international cooperation started and ended with the Soviet Union. The communist superpower had been the first to put a man into orbit in 1961 and had been on a par with the US for much of the decade in their attempt to put a man on the moon. In many ways, a partnership with the Soviets now made good sense. "I decided, and I hope I made the right decision, that although Jim Webb certainly had done a tremendous job of building up NASA and the program on the basis of the Russian threat, that times had changed," Paine reasoned. "The time had come for NASA to stop waving the Russian flag and to begin to justify our programs on a more fundamental basis than competition with the Soviets."[13]

The Soviet Union had long been the United States' top priority in international relations. This dated back to the end of World War II, as both sides fought for control of German rocket scientists, their intellectual property, their research, and the rockets that they had built during the war.[14] In the ensuing years, the secretive nature of its space program was a microcosm of the closed communist state. From its inception, the space age had provided both sides with the chance to demonstrate their differences to the world. The sense of national resolve, the strengths of each government, the differences between communism and capitalism, and the vision of their leaders—all factored into the Space Race. Paine often pointed out that space exploration was a major stabilizing force during the Cold War. It provided a peaceful alternative to direct military confrontation. The world's two superpowers competed openly without upsetting the delicate balance of international power.[15]

East-West cooperation changed drastically following Apollo. After the wake-up call of *Sputnik*, the United States had approached the Soviet Union about cooperating in the new field of space exploration. The response had always been that "we could not cooperate until there was complete and total disarmament."[16]

As early as 1959, President Eisenhower had offered to help track Soviet satellites. The US was then building a new network of global ground stations called the Spacecraft Tracking and Data Acquisition Network. The reply was always no. President Kennedy proposed in 1962 to exchange satellite tracking and science data; this too was rejected. During NASA's Project Gemini, President Johnson invited Soviet officials to attend the launches. All were dismissed by the Politburo. It got to the point where the Soviets stopped answering letters.

A breakthrough of sorts came with the Dryden-Blagonravov talks. Between 1962 and 1965, agreements were reached for four projects in meteorology, communications, geomagnetic surveying, and space medicine. But the progress was slow, and by all accounts, disappointing. NASA had also tried to approach the Soviets through the United Nations Outer Space Committee and the International Committee on Space Research (COSPAR). Neither proved particularly helpful. Most of the time, there was no response. Meaningful communication had become practically nonexistent by the time Paine arrived at the agency.

The refusal to cooperate was not so much due to differences over science as it was dictated by Cold War ideology. Paine once explained why he thought it was so difficult to bring the Soviets to the negotiating table. He pointed specifically to the different types of men at the top who were making the decisions. Most senators and congressmen in Washington, having been lawyers at one point, put their trust in the process of litigation, compromise, and cooperation to reach a common goal. Most in the Soviet Politburo were trained in the military, where cooperation was seen as a sign of weakness.[17] On another occasion, he said:

> There is nothing wrong with nationalism. It has great advantages, and a couple of major disadvantages—war being the chief among them. But I think when you start looking at a planetary movement of our species . . . to do this on a nationalistic basis does not make any sense at all. It is a movement of all mankind, and we ought to do it on a planet-wide basis. Now, I argue that America should lead. I believe we have the highest technology in the world, the strongest economy, a frontier tradition. We also possess the richest diversity of ethnic groups in the world . . . and for that reason I think we are the logical people to make sure that when we move out to these new worlds, we take with us all of the cultural diversity of mankind.[18]

Leading up to Apollo 11, he tried to reopen the dialogue with the Soviet Union. He sent a series of letters to academician Anatoli A. Blagonravov. On April 30, 1969, he put in the mail a copy of the agency's new publication, "Opportunities for Participation in Space Flight Investigations," to Blagonravov. Paine assured him that any reasonable proposal would be welcomed. He would review them himself. A month later, he invited Blagonravov to see the launch of Apollo 11. Blagonravov wired back that he would not be attending.

Thirty thousand feet above the Pacific, Tom Paine, President Richard Nixon, National Security Advisor Henry Kissinger, and Secretary of State William Rogers gathered in the staff lounge of the Boeing 707 for some coffee. *Air Force One* was heading west on the last leg of its flight to Hawaii for the splashdown of Apollo 11. The four spoke candidly and at length about the defining moment of the nation's space program that would soon be in the books with the safe return of Armstrong, Collins, and Aldrin from the moon. Now, Nixon wanted to use that victory as a springboard for reaching out to the Soviet Union. He asked Paine to refocus his effort so that an agreement could be reached on a cooperative space program. It was not the first time that the administration had looked in that direction. But those occasions paled compared to this, a watershed moment with Paine and a private, captive audience with the president and his right-hand advisers. As they talked, he agreed that the timing was right, but cautioned the president, if nothing more than to downplay the expectations, that any progress would almost certainly be slow.[19]

When he returned to Washington (he did not accompany Armstrong, Collins, and Aldrin on their "Giant Leap" celebratory world tour), he invited Mstislav Keldysh, president of the USSR Academy of Sciences, to a September meeting at NASA Headquarters. "The Chief Theoretician," as he was known in the West, said that he was open to a discussion. NASA was just then evaluating proposals for experiments to go onboard with the Viking Mars robotic landing missions scheduled for 1975.[20] Paine mainly wanted to see if there was any interest on the Soviets' part in participating with the United States in planetary exploration. Upon receiving it, Keldysh wrote back and courteously declined to attend, saying there was just not enough time to make travel arrangements for him and his scientists. But he was more terse about the Mars proposal. He told Paine they were not interested in putting experiments on the Viking landers. The Academy of Sciences instead suggested a more modest proposal. Perhaps the two sides could coordinate planetary goals and exchange results from deep-space robotic investigations. Although Keldysh had declined to attend the meeting, he asked for copies of the material for his scientists to review. Paine told Kissinger that it was a small but somewhat encouraging step. This time, he at least got a response from the Russians.

He sat down and wrote another letter to Keldysh in October. This time, he sent along a copy of a report by president Nixon's Space Task Group (STG) on plans for the future of the US space program. His move divulged to the Soviet Union, in full, the plans of the United States in space for the next decade. In return, he specifically wanted to gauge the potential for the two sides to carry out a joint manned mission that would lead to "major complementary tasks to benefit both our countries."[21] The STG had concluded that, for the first time, it might be possible to use the space program to advance the mutual political relationships of the two countries. Keldysh agreed, but wrote Paine that he again wanted to defer further discussion on the specifics as to the time and place for a meeting for another three months, until the end of the year.[22]

A serious proposal for some sort of joint US-USSR mission (either manned or unmanned) was now taking shape. For the first time, the hard-line Communist Party first secretary, Leonid Brezhnev, conceded from Moscow that the time had perhaps come to collaborate in space. Paine was delighted. At first, he told Mueller to consider a launch exchange mission in which each side would launch a spacecraft from the other. The suggestion was quickly dismissed by his own managers, however, on technical grounds. Adapting a Soyuz spacecraft to the Saturn IB launch vehicle would have been prohibitively challenging and risky. The same would have been true of putting an Apollo spacecraft on top of a Soviet rocket. Managers in Houston were correct when they predicted that the US would not be able to gain access to enough technical data to satisfy engineering and safety requirements.[23]

He had another idea, which he discussed with George Low. Perhaps a Soyuz spacecraft could rendezvous and dock with the American Skylab orbital workshop (then scheduled for launch in 1973). This was an idea that Olin E. Teague, then chairman of the House Committee on Science and Astronautics' Subcommittee on Manned Space Flight, had approached him with. The longtime congressman from Texas was a fierce defender of NASA. One day over lunch after a hearing, he had suggested to Paine that a coordinated emergency response using the American space laboratory might work. The idea was feasible technically, as the multiple docking adaptor on Skylab had two docking ports. It could, therefore, accommodate two spacecraft at the same time.

But Paine's motivation was not technical so much as a desire on his part to let Keldysh know that he was serious. He knew that the proposal, in all likelihood, had no chance. It was too ambitious for what the Soviets wanted. The Skylab schedule was already slipping. Nevertheless, he proposed it to Keldysh. Nothing materialized, but it kept the dialogue going. If the Soviets did not like the idea (which his own people in the Skylab Program Office at NASA were, as he had predicted, strongly against), perhaps they could suggest something different. At least the talks would continue.[24]

With the White House behind him, he was ever confident that the US could build a joint manned Mars program from the effort. This was his ultimate aim. The exchange of letters with Keldysh picked up pace in the spring of 1970. Philip Handler, president of the National Academy of Sciences, also opened a series of informal talks with Soviet scientists in the summer of 1970 that turned out to be pivotal. The tedious negotiations tested Paine's diplomatic patience, however. Arnold Frutkin found it equally unproductive. While Keldysh and the academic and technical establishment started to show authentic signs of interest and a general willingness to proceed, the party leadership in Moscow was simply not yet ready to commit to an agreement with Washington.[25]

The telephone in his office was ringing; it was late in the day. On the other end was Phillip Handler from the National Academy of Sciences. Four days earlier, Soviet Ambassador to the United States Anatoli Dobrynin had asked Handler to make an appointment with the newly appointed science attaché at the Soviet embassy, Evgeniy Belov. He hinted that Belov had brought with him a significant development from Moscow that he needed to discuss with him.

In the meeting, Belov had told Handler that the Academy of Sciences was prepared to formally discuss "matters of mutual interest." Surprised, Paine immediately got on the phone and called Frutkin at home to move quickly and draft a letter to Keldysh agreeing to establish a joint project. His reply initially revolved around not any one specific mission, but rather, the wider topic of international space rescue and development of a common docking mechanism that could be developed to allow a Soviet spacecraft in distress to link up with an American spacecraft. It was not a new idea, but now it was actually happening. Managers in Houston and Washington had been discussing a mission along those lines for months. Two months earlier, in May, he had used the analogy in a speech at the Apollo History Workshop in Washington that ships of all nations perform rescues at sea. While not the Mars mission that he had hoped for, it was the kind of specific response that up until then had been entirely missing from the talks.[26]

The Soviets followed with a request for a general agreement. They wanted a meeting between engineers so that design of a "mutually acceptable, common space docking mechanism" could begin. Such a capability would provide both sides with the ability to return astronauts or cosmonauts to Earth in case of an emergency. A week later, word of the still-private talks leaked out into the open during the International Astronautical Federation Congress in Konstanz, Germany. The Soviets moved to grab the headlines. Keldysh heralded the talks they had started as "in the interest of world science and the progress of all mankind." Paine saw it for what it really was: the first tangible step of many that were still needed toward real cooperation—cooperation that the US had initiated. He invited Keldysh to send a delegation as early as practical to the Manned Spacecraft Center to begin work. This time, the invitation was accepted.[27]

Technical discussions in Moscow, Washington, and Houston ensued over the next two years. On the US side, Low and a delegation from Houston led by former Apollo flight director Glynn S. Lunney worked out the specifics of a joint proposal.[28] The team finalized a seventeen-point technical agreement in Moscow in April of 1972. On May 24, President Nixon and Prime Minister Kosygin signed the pact under bright lights and cameras. A joint manned mission to low-Earth orbit was now official. NASA gave it the name Apollo-Soyuz Test Project (ASTP).

The groundwork was laid. Now, both sides could begin work on the technical aspects of the world's first international spaceflight: the common docking module, a joint flight plan, a complex orbital rendezvous, the language barrier, and crew training. ASTP flew three years later, in July 1975. The fact that it was carried out using mostly US technology did not deter either side from claiming victory, mostly of the political variety. The highly anticipated flight was the most-watched space mission in history. People from the Eastern Bloc nations and in the West tuned in to see the rocket launchings on live television, the first time a rocket launch was broadcast live to the Soviet people. Three astronauts and two cosmonauts spent two days docked in LEO as Mission Control in Houston and Baikonur jointly directed the flight from the ground.[29] Paine had left NASA by then. Before the mission, new NASA Administrator James C. Fletcher wrote to Paine and thanked him for his persistence and perseverance, which had made it possible for ASTP to happen. The letter was uncommonly laudatory, declaring that US relations with the Soviet Union had improved "in an almost spectacular way" on his watch.

Paine read the letter and quietly filed it away.[30]

Tom Paine. (NASA)

Lieutenant George Paine and two-year-old "darling daddy boy" Tommy. (Library of Congress)

Childhood home during the Great Depression. (Library of Congress)

Mathew Fontaine Maury High School in Norfolk, Virginia, where Paine graduated with the class of 1938. (Norfolk Public Schools)

USS *Pompon* off the coast of Brisbane, Australia, in July 1943. (US Navy)

Mare Island Naval Shipyard near San Francisco, where young Tom Paine caught his first glimpse of fleet submarines and warships. (US Navy)

SOUTHWEST PACIFIC 1943

Tom Paine's wartime journal. (Library of Congress)

ABOARD an American submarine tender, men of the U.S. Navy get lessons in the Japanese names for equipment on submarines so that they can man the boats which remain in the surrendered Japanese fleet. U.S. Navy Photo

A newspaper photograph showing officers aboard the USS *Euryale* being briefed on the captured submarines of the Imperial Japanese Navy in the fall of 1945. Tom Paine can be seen near the left edge of the picture by the hand drawn arrow and the word "ME." (US Navy)

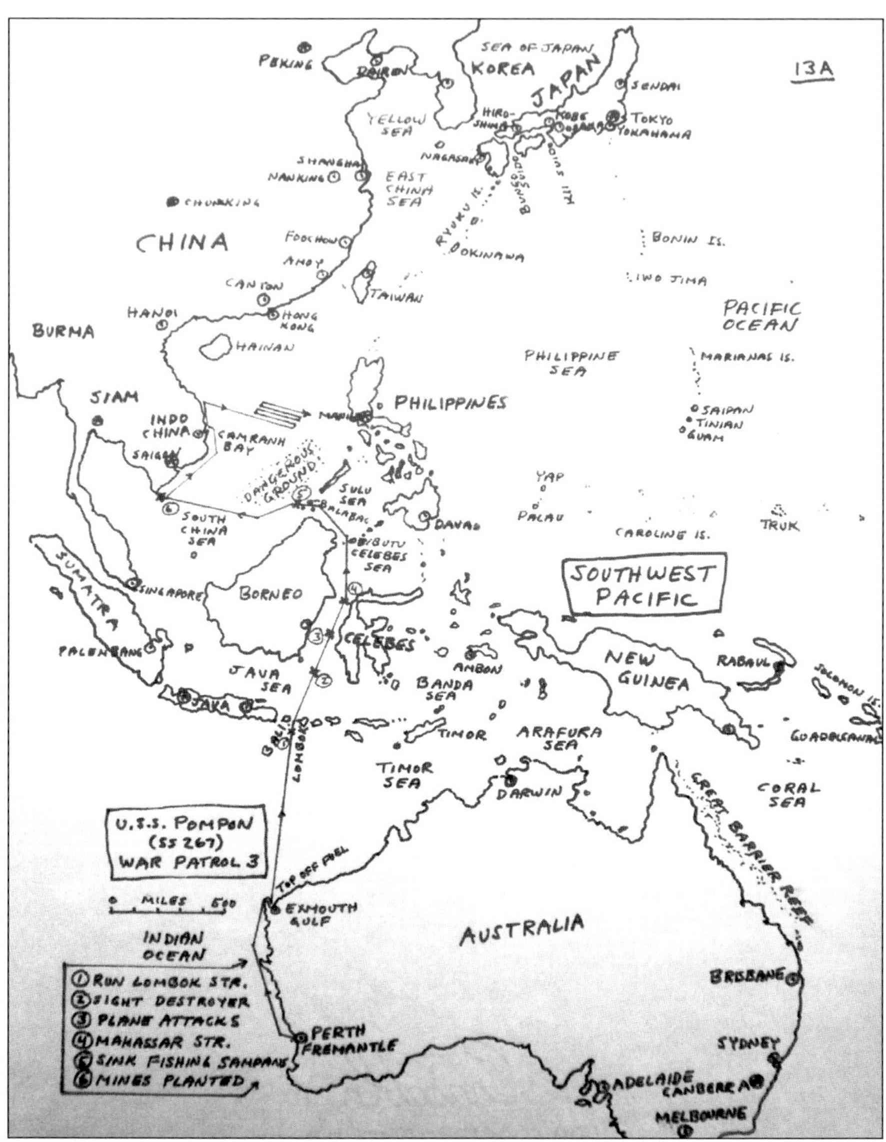

Paine's first action in the Pacific was on the third war patrol of the USS *Pompon* during which she sank two Japanese patrol boats. This map that he drew shows the submarine's patrol area from November 1943 to January 1944. (Library of Congress)

United States Pacific Fleet

Flagship of the Commander-in-Chief

The Commander in Chief, United States Pacific Fleet, takes pleasure in commending

LIEUTENANT (JG) THOMAS O. PAINE
UNITED STATES NAVAL RESERVE

for service as set forth in the following

CITATION:

"For distinguishing himself by meritorious conduct in action in the performance of his duties in a United States Submarine during a war patrol of that vessel. His precise control of depth was of valuable assistance to his Commanding Officer in conducting attacks which resulted in sinking more than 4,000 tons of enemy shipping, and damaging over 7,000 tons. In addition he voluntarily went over the side in extremely cold water to plug a sea connection. His calm manner and devotion to duty contributed directly to the success of his vessel. His conduct throughout was an inspiration to the officers and men in his ship and in keeping with the highest traditions of the United States Naval Service."

Commendation Ribbon Authorized

C. W. Nimitz

C. W. NIMITZ,
Fleet Admiral, U.S. Navy.

Paine received the Pacific Fleet Commendation Ribbon with Combat Clasp for Performance in Action for meritorious conduct on the night of August 9, 1944. (Library of Congress)

USS *Euryale* with captured Japanese submarines I-401, I-14, and the I-400 in Sasebo Harbor on November 16, 1945, preparing for the transpacific journey to Pearl Harbor. (US Navy)

The USS *Barb* (SS-220), *Pipefish* (SS-388), and *Pompon* (SS-267) arrived back at Pearl Harbor on the same day following the war. (US Navy)

Tom and Barbara Paine cut their wedding cake on October 1, 1946, in a traditional naval ceremony at the Long Beach Naval Shipyard. (Library of Congress)

Working the stress-rupture machine as a graduate student in 1948 in the Stanford University School of Mineral Sciences laboratory. (Library of Congress)

A page from "Schenectady Works Welcomes You!" published by General Electric in 1949, the same year that Paine began work as a materials engineer. Pictured on the page was a 1928 Thomas Edison commemorative plaque and an aerial view of the "City Within a City." (General Electric; Schenectady County Public Library)

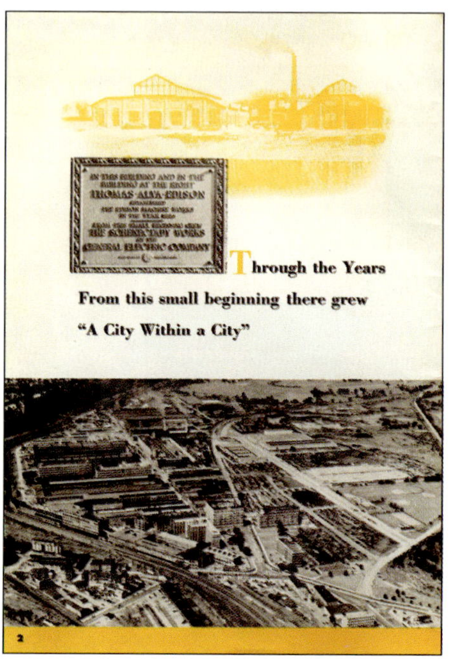

A lighter moment at the General Electric materials laboratory in Schenectady. Paine is third from the right. (Library of Congress)

Addressing a gathering as the general manager of GE TEMPO in 1963. (Library of Congress)

NASA Headquarters was located in three buildings near the United States Capitol in Washington, DC. Paine's administrator's office was in Federal Office Building Six, located at 400 Maryland Avenue, SW. (NASA)

The iconic triad of NASA Administrator James Webb (center), Deputy Administrator Hugh Dryden (left), and Associate Administrator Robert Seamans in 1966. (NASA)

The newly appointed deputy administrator of NASA in the spring of 1968. (NASA)

The first seven manned Apollo flights flew with Tom Paine as the head of NASA. Apollo 8 was the first to orbit the moon, and Apollo 11 was the first to land on the moon. Each crew designed their own mission emblem. (NASA)

Touring a mockup of the Skylab Orbital Workshop at the Marshall Space Flight Center in Huntsville, Alabama, in April 1968. From left to right are William Brooksbank of the Marshall Space Flight Center, Paine, Wernher von Braun, Clare F. Farley from NASA's Office of the Administrator, and Charles J. Donlan of Marshall. Paine is holding an ordinary man's shoe outfitted with a special anchoring device in the sole to hold an astronaut to the floor of the workshop in the weightlessness of space. (NASA)

Robert Gilruth, director of the Manned Spacecraft Center, greets Paine in Houston in April 1968. (NASA)

President Johnson presents Apollo 7 commander Wally Schirra with the NASA Exceptional Service Medal on November 2, 1968, following the mission. Paine was NASA's acting administrator at the time. (NASA)

Paine addresses the crowd in New York's City Hall Plaza after a presentation ceremony honoring the crew of Apollo 8 on January 10, 1969. On the right is Barbara and on the left is mission commander Frank Borman. (NASA)

BB/

Reproduced at the Richard Nixon Presidential Library

FG 164/A

The White House,

MAR 6 - 1969 19

To the

Senate of the United States.

I nominate Thomas O. Paine, of California, to be

Administrator of the National Aeronautics and Space Administration.

RICHARD NIXON

Paine's nomination letter from President Nixon to the US Congress. (Nixon Presidential Library)

President Richard Nixon announces Paine's appointment as NASA administrator on March 5, 1969, at the White House with Vice President Spiro Agnew looking on. (NASA)

Paine makes a brief statement after being sworn in as the third administrator of NASA in Vice President Agnew's office on April 3, 1969. In the back are George, Judy, Barbara, and Frank. (NASA)

Paine talks with Vice President Agnew as they await the launch of Apollo 11 at the Kennedy Space Center. At center is Apollo 8 astronaut Bill Anders and at left is Lee James from the Marshall Space Flight Center. (NASA)

A replica of the plaque attached to the descent stage of the lunar module that was left on the moon commemorating the first lunar landing. (NASA)

Paine and Nixon on the bridge of the *Hornet* awaiting the splashdown of Apollo 11 in the early hours of July 24, 1969. (NASA)

The motorcade celebrating Apollo 11 in Manhattan on August 13, 1969, drew a crowd estimated at four million. Seated in front of Buzz Aldrin is a beaming Paine. (NASA)

Decision meeting at the Cape on April 10, 1970, on whether to proceed with the launch of Apollo 13. Standing at the head of the table is NASA's Director of Medical Research and Operations Dr. Charles Berry. Seated counterclockwise are Paine, Rocco Petrone, Julian Scheer, Dale Myers, George Low, and Chester Lee. (NASA)

Paine shields First Lady Pat Nixon from the rain while the president, Tricia Nixon, and Barbara watch the Apollo 12 countdown. It was the first time that a sitting president witnessed a NASA launch. (NASA)

Paine discusses the revised Apollo 13 flight plan with the president. With him are Rocco Petrone and Chester Lee. (NASA)

Following the launch countdown of Apollo 13 in Firing Room One at the Cape. (NASA)

Swigert, Lovell, and Haise talking with Paine aboard *Air Force One* on April 19, 1970, returning to Houston. (NASA)

Neil Armstrong is sworn in at NASA Headquarters on July 1, 1970, to head the agency's aeronautics programs as the deputy associate administrator for aeronautics. (NASA)

Jim Lovell relates to the members of the Senate Space Committee in an open session on the problems of the Apollo 13 mission. Rocco Petrone can be seen behind Paine. (NASA)

Vikran A. Sarabhai, chairman of the Indian Space Research Organization and head of India's Department of Atomic Energy, signs a joint satellite agreement with the United States on September 18, 1969. (NASA)

Secretary of State William P. Rogers signs the guest register at the Kennedy Space Center on April 11, 1970, as Paine and Willy Brandt, chancellor of the Republic of Germany, look on. (NASA)

Spiro Agnew presides over a meeting of the National Aeronautics and Space Council on October 28, 1969. From left are: Presidential Science Advisor Lee Dubridge, Chairman of the Atomic Energy Commission Glenn Seaborg, Under Secretary of State for Political Affairs U. Alexis Johnson, Agnew, Council Executive Secretary Bill Anders, Secretary of the Air Force Bob Seamans, Paine, and Assistant Director of the Bureau of the Budget James Schlesinger. (NASA)

Station Director Tom Reid giving Paine a tour of Deep Space Station DSS-42 at Tidbinbilla in the Australian Capital Territory on February 24, 1970. (Colin Mackellar; Clive Jones and Mike Dinn)

A rendition of what a 1990s mission armada to Mars might entail as envisioned in 1976 by NASA. Overhead are several solar sail-powered interplanetary shuttles each carrying up to 25 tons of cargo. Two crew capsules have landed, as have various rovers and processors to mine and study the Martian surface. A nuclear power station preceded astronauts to the surface. According to the space agency, sustained research and development could make such a three-year mission by a six-person crew possible by the mid-1990s. (NASA)

Logo ideas for the National Commission on Space. "It must be simple for clarity in small size," said Paine. "My candidate is the upper-right." That was the one the commission chose. (Laurel Wilkening)

Visit to the Hitachi Mito Works in Hitachinaka, Ibaraki, Japan on June 29, 1972. Standing with Paine and Hitachi officials in front of the 90-meter elevator test tower are Ed Schmidt and George Grega, CEO of General Electric Japan. (Library of Congress)

President Reagan's Science Advisor Jay Keyworth introduces the National Commission on Space at a White House press conference on March 29, 1985. From left: Tom Paine, Luis Alvarez, Paul Coleman, George Field, William Fitch, Keyworth, Charles Herzfeld, Jack Kerrebrock, Jeane Kirkpatrick, Kathryn Sullivan, David Webb, and Laurel Wilkening. Not present were Neil Armstrong, Gerard O'Neill, Bernard Schriever, and Chuck Yeager. (White House)

An artist's rendition from 1975 of a toroidal space colony designed by NASA's Ames Research Center. (NASA)

Paine and Wilkening present *Pioneering the Space Frontier* to President Reagan on July 22, 1986. (The White House)

The Augustine Committee. Front row left to right: Pete Aldridge, Norm Augustine, Laurel Wilkening, Tom Paine. Back row left to right: Robert Herres, David Kearns, Daniel Fink, Don Fuqua, Louis Lanzerotti, D. James Baker, Edward Boland, and Joe Allen. (NASA)

Dean Francis B. Sayre, Jr. blesses the Space Window in the Washington Cathedral in the dedication ceremony on July 21, 1974. In the background: Mrs. And Mr. Rodney Winfield, James Fletcher, Michael Collins, Neil Armstrong, Buzz Aldrin, and Tom Paine. (NASA)

Official NASA administrator's portrait by William Draper. (NASA)

10

WHAT NOW?

The preferred configuration which is emerging from these studies is a two-stage delta wing reusable system in which the orbiter has external tanks that can be jettisoned.
—James Fletcher, describing the planned space shuttle in June of 1971

The space shuttle models looked like something from a science-fiction movie. Machined from polished stainless steel, they were shiny and precise. He picked one up. George Mueller told him that particular design was being proposed by the Boeing Company. More spaceplane than airplane, the scaled replica of the next generation of spacecraft did not look anything like the familiar rockets and capsules that the US, Soviet Union, and the rest of the world had been launching into space when the end of World War II had ushered in the space age. All the shuttle designs looked quite esoteric to him.

He had just walked in the snow across Sixth Street to Federal Office Building 10B, where Mueller's office was. Coming in from the cold, he sat down with a cup of coffee and continued the conversation that he had started with Mueller on the phone. It was late in the fall of 1969, and Paine was meeting with his associate administrator for manned space flight. The first manned landing on the moon was now behind them. The biggest question facing them now was, what was next?

At $3.3 billion, NASA's budget was neither bold nor balanced. It was, in fact, the agency's lowest level of funding since 1961, and more than half a billion dollars below what Congress had appropriated the year before. The agency's post-Apollo plans were ill-defined mainly because it was underfunded. Congress had made across-the-board budget cuts. Nothing was spared, and NASA was one of the agencies hit hardest. Forty-five thousand layoffs that year had the aerospace industry as a whole hurting badly.

The situation at NASA was not promising. The draconian cuts affected more than just a few programs:

- The Mississippi Test Facility, on the banks of the Mississippi-Louisiana border, had been placed on standby status after having completed the final rocket engine test of the Saturn V.
- Saturn V launch vehicle production was now suspended; no more moon rockets would be made.
- Apollo 20 was canceled, along with the production of all Apollo spacecraft hardware.
- The launch of Skylab, the Earth-orbiting workshop, was postponed.
- Apollo 18 and 19 were now deferred until 1974, after the Skylab missions. They would be canceled altogether in September 1970.
- Launch of the highly anticipated robotic Viking Mars landing missions was postponed from the 1973 to the 1975 launch window.
- NASA's Electronics Research Center in Cambridge, Massachusetts, was shut down and later turned over to the Department of Transportation.[1]

Tom Paine was now fighting hard not to lose the remaining three Apollo flights (Apollo 15, 16, and 17). In no small twist of irony, the scientists who for years had deplored the human spaceflight program were now leading the charge to keep Apollo alive. They still wanted a more extensive exploration of the moon's surface. There were still many unanswered questions about the moon, questions on its origin, mineralogical makeup, the existence of water and proto-biological materials, and soil characteristics. Most important was the question as to whether or not the surface was suitable for mining.[2]

Since arriving at NASA, Paine's critics had contended that he did not make organized scientific research a priority. NASA was supposed to be a scientific organization, they argued. He openly admitted that he saw the role of his agency differently. NASA was not the "czar of science," deciding on the meaning of scientific values. That was the mission of the universities and the academics. He maintained that while large organizations like NASA were needed to take on big challenges like space exploration, the foundation of science "is and must be the individual scientist."[3]

Paine was an engineer caught up in a complex infrastructure populated by a lot of scientists. He was, however, quite cognizant of the differences between the two. Engineers, he espoused, must work together as a team to build a bridge, construct an offshore oil rig, or get a five-hundred-ton airliner up into the air. They succeeded as a team or failed as a team. But while the engineer knows he cannot do it alone, the scientist can often afford to think differently and still be very successful. He can be more individualistic, more eccentric, and more theoretical. But since NASA was trying to get to the moon and beyond, engineering had to take priority over science.

The debate over engineering versus science escalated as never before during the Apollo program. A case in point was the elimination of the Apollo Lunar Surface Experiment Package, a suite of scientific equipment and experiments from the first moon landing that was dropped primarily for reasons of weight. "It turned out to be a good thing," Paine said in hindsight. "Armstrong and Aldrin had enough to do. What more could a scientist-astronaut have done?"[4] For him, the chief purpose of the nation's space program was nothing if not the epic human exploration of outer space.

Apollo was epochal because men, and not machines, went to the moon. Whether people should go into space when robotic probes might suffice had been the subject of debate from the earliest days of the space age. The US had to decide very early on whether it should develop the capability for humans to operate in space or leave this area to the Soviet Union. But the

Soviets' intention was clear: They wanted cosmonauts in space. That had ended the debate, giving NASA a charter and a purpose that allowed the US to eventually succeed in space.

But to send people into outer space, the US had to develop some very advanced technologies. It was an expensive proposition; almost 50 cents on every dollar of the space budget went to human spaceflight. Vocal opposition came from well-respected figures in the space program. Bruce Murray, a cofounder of the Planetary Society, said at the time that a man in space was only meaningful when he was needed for psychological reasons. James Van Allen, discoverer of the radiation belts around Earth that bear his name, declared that the "vaunted advantages" of having human crews in space was "rubbish." Even President Nixon's own science advisor, Lee DuBridge, was never fully convinced that human spaceflight was needed for what the country was trying to accomplish in outer space.[5]

Keeping a balanced space program was important to Paine, but he fully conceded that the success of human flight had to come at the cost of other programs. Recognizing that unmanned programs generally could not evoke the kind of enthusiasm, romance, and deep emotion that manned programs brought, he did not waver in his position. By doing so, he kept the rift open between him and the scientists. (In Houston, Gilruth and Kraft wanted to know the absolute minimum scientific requirement they could get away with on a moon flight.) "It was very difficult to keep a positive constituency among the scientists. . . . I think some of them adopted a 'rule-or-ruin' position: either we did what they said or they would publicly attack us, and then they *would*. It wasn't an idle threat."[6]

As the 1960s ended and the calendar turned to a new decade, many pressing issues confronted America. There was the ongoing conflict in Southeast Asia, inflation, fiscal instability, and the beginning of the first global energy crisis. All placed competing fiscal demands on government. For most of the decade, Project Apollo had been carried out with success and safety as prerogatives. Cost was secondary, at least in the beginning. But when the time came to decide what would come after Apollo, cost drove everything.[7]

The United States in 1970 was in no rush to construct a space station. In 1961, when Webb presented Kennedy with options for the United States in outer space, the president chose to go with a moon landing. The decision was bold but not the riskiest. It was , however, calculated enough to mobilize a generation of aerospace workers and capture international attention. Fail or succeed, it would put America front and center on the world stage. Had he elected, instead, to build a space station, the US would presumably have had, by 1970, a large, permanent habitat of some sort circling Earth.[8]

Paine did not agree with his top advisers on what the next big program should be. He wanted to proceed directly to a moon settlement and a Mars mission, but understood that the timing was wrong. It would not have been politically or fiscally tenable. Others, like Edgar M. Cortright, the director of the Langley Research Center in Virginia, lobbied for a large Earth-orbiting laboratory that could serve both the science and military communities.[9]

But NASA had no funding or mandate to construct a fantastic "station campus," as Wernher von Braun called it. From the Marshall Space Flight Center in Huntsville, he talked to Paine about building a "spacebase." It would be quite large, big enough to accommodate fifty or perhaps even a hundred persons. Arthur C. Clarke's *2001: A Space Odyssey* had brought the concept into full living color. Space enthusiasts believed that a giant wheel turning gracefully in the sky could serve as a gateway to the moon and the planets. It was not only believable but plausible. Paine knew Clarke, and was very much taken in by his friend's visionary prognostications. But in order to get there, an affordable stopgap measure called Skylab was first needed.

Skylab provided the US with its first Earth-orbiting laboratory. Scientists wanted to know how the human body would respond to the prolonged exposure of living and working in space. The largest piece of Skylab was the Orbital Workshop. The spacious 12,400-cubic-foot living module was converted from the S-IVB third stage of a Saturn V rocket. It could accommodate three astronauts in space for up to three months at a time. In October 1968, Paine had made a key decision as the acting administrator of NASA by directing the agency to proceed with Skylab. It was the first project to fly in the so-called Apollo Applications Program.[10] From May 1973 to February 1974, three crews would spend a total of

171 days in orbit. The agency wanted to use Skylab for a few years before building a permanent spacebase. That would come later, after the space shuttle was operational.[11]

He clearly believed that the conquest of space was of such enormity that the United States could not and should not tackle it alone. Speaking in Canberra, Australia, he used Antarctica as an analogy. Paine proposed that the international focus of a space station should be like exploring the icy continent of Antarctica. Like Antarctica, it would be a place where various nations would operate a mix of science stations. Nations would cooperate with one another there to support the mission at large. At the time, the international aspect of a space station was still very poorly defined. The details of the program itself were nebulous. He advocated that while the United States must be the standard-bearer, foreign nations should be invited to participate. Many countries could use it as a shared resource. But NASA had no plans to internationalize the station in terms of ownership, he said in 1970.[12]

He had written to the president in February 1969, saying that Nixon now had "a unique opportunity for leadership that will clearly identify your administration with the establishment of the nation's major goals in spaceflight for the next decade." Paine started out very boldly with Nixon. Dismissing what the Bureau of the Budget had to say, he urged the president to pursue a clearly defined path for America in space. A permanent space station was the first step. A settlement on the moon was next. Human exploration of Mars would complete the equation.[13]

The recommendation was premature, however. Nixon wanted more options. He wanted something that was less expensive, with more immediate returns, and frankly, not so ambitious. Always thinking politically and already with an eye toward the next election, he wanted a program that would quickly create new jobs for the aerospace industry in key congressional districts.[14] On February 13, 1969, he appointed Vice President Agnew to chair the administration's Space Task Group.[15] Their charter was to come up with a plan for the future of the US in outer space for the upcoming decade.

Nixon knew that Agnew, like Paine, favored a large space program. He was counting on his science advisor, Lee DuBridge, and to a somewhat lesser degree, Robert Seamans (representing the secretary of defense), among others, to bring some balance to the group. The State Department, Atomic Energy Commission, Budget Bureau, and Congress all had representatives on the STG. Many issues needed to be made clearer. They affected both the military and the civilian space programs. In a nutshell, the STG's job was to map out a strategy so that politically tenable, flagship missions could be implemented in a Nixon space policy. Nixon told Agnew to have the report on his desk by the first of September.[16]

The STG had a very difficult job. Seamans openly questioned the practicality of Paine's Mars agenda from the beginning. The two were friends, on good terms, and usually supported each other's positions. But his time away from NASA had convinced Seamans that there was actually little public support for something as lofty as a master plan to establish a permanent human presence in outer space. In his own words, he was "quite at odds" with Paine (and with Agnew).[17] He also found, much to his surprise, that even DuBridge, who was usually lukewarm to any calls for a grandiose human space program, wanted to keep the Mars option open.

With less than a month before the report was due, time was running short. Coming out of a particularly heated meeting on August 4, Seamans had had enough. He talked the situation over with his boss, Melvin Laird (Nixon's secretary of defense), and wrote a letter to the president to explain his position. He urged strongly that the US not commit itself to a manned Mars mission. He also recommended to the president that the country hold off, at least for the time being, developing a space station. Instead, he advocated for a National Space Transportation System, which by all accounts should greatly reduce the cost per pound to launch payloads into orbit for both NASA and the military. As the secretary of the Air Force, it was the most sensible case he could make for the Defense Department. Seamans's action disappointed Paine when he found out what he had done. He thought his friend was out of line in circumventing the council and going directly to Nixon, and lamented that "we should have recommended to the president a real gung-ho civilian space program for the 1970s." He was starting to see the writing on the wall.[18]

Paine also wanted the space shuttle, but merely as part of a much broader plan to build and then service a permanent, large-scale space station. He went on record several times in the STG to say that the station and the shuttle should be built at the same time, and presented numbers that showed it could in fact be done. All that was needed was to restore the agency's funding to the level that it had during Apollo, at 1 percent of the gross national product.[19] He envisioned the finished space station then as a gateway to launch expeditions to the moon and Mars. The space shuttle would have been only one component (in fact, the smallest component) of the whole plan.

But Paine knew Nixon, and knew that space exploration was not a priority for him. He had no choice, however, but to operate within the confines of the STG. The president got the report, and on March 7, 1970, a Saturday, presented his plan for the US in space to Congress. He asked that the country develop a reusable space shuttle as its next flagship program in space. This was to be followed by a space station at some later time. Paine had lost. Most disappointing to him was the part of the plan that called for the US to wait on a Mars mission until the year 2000 (thirty years in the future). By deferring a Mars landing to the end of the century, Nixon did not have to address the issue at all. Paine's inability to win over the White House also doomed the space station.[20] The recommendation that the White House went with was the most conservative and least ambitious of all the options that the group had considered. With his hands tied, Paine had little choice but to go along with the recommendation if NASA wanted to continue at all with human spaceflight. Nixon really had no interest in the space program beyond the acclaim that Apollo brought. While he would not kill the space program, he did not have to be lavish with it either. Paine's one great failure as the administrator of NASA meant that there would be no gateway in low-Earth orbit that astronauts could use to go back to the moon and on to the Red Planet in the foreseeable future.[21]

He did not like the decision, but the go-ahead for the space shuttle meant that NASA at least had its flagship program for the decade of the 1970s. In 1970, predictions of what the space shuttle could do ranged from the sanguine to the fantastic. Most space insiders presumed that a fleet of reusable shuttles would provide easy and routine access into space. In a December 1969 *Air Force and Space Digest* article, the space shuttle was hailed by the United States Air Force as the key that would open the doors of space just as the intercontinental railroad opened the American West for the common man. Astronauts would be as numerous as airline pilots. Flights into orbit would be scheduled almost weekly; virtually any healthy person who could afford to travel into space could do so.[22]

Paine echoed the seemingly ubiquitous thinking at the time that access to space would be very routine using the space shuttle. His high expectations regarding what the shuttle might do were primarily pragmatic. To him, the shuttle was a means to an end, a way to access LEO cheaply. "We are not interested in this as a Buck Rogers technique," he said. "Our principal interest is the fact that this promises to reduce by a factor of 10 the costs of operating in space."[23]

His enthusiasm was based in no small part on initial technical studies done at the time. Competing aerospace giants Boeing, Convair, Lockheed, Martin Marietta, McDonnell Douglas, and North American Rockwell all eagerly and rightly wanted the shuttle prime contract. Their competing numbers all showed very low cost projections to operate the spaceplane, which was not surprising to anyone. Industry reports boasted that it might even be possible to do better than the goal of $100 per pound-to-LEO price tag that NASA and its contractors had projected (a figure that the space program is nowhere close to even today). If that goal were achieved, the space shuttle could have replaced, based on cost alone, all of the mid-to-heavy expendable launch vehicles that were then in production.

The design also became tied to the Department of Defense (specifically the US Air Force). For funding appropriations to be approved, Congress had stipulated that the shuttle be used by both NASA and the military. Paine and Seamans signed an agreement in December of 1968 for dual use of any shuttle that NASA would develop. The Air Force had equal ownership in the program. The final wing design of the space shuttle and the size of the payload bay were, in fact, dictated not by NASA but by Air Force requirements. Like NASA, it planned to launch its own fleet of shuttles into orbit. A massive launch complex at Vandenberg Air Force Base on the coast of California would launch dozens of missions a year into polar orbit around Earth.[24] Paine signed the agreement believing that the synergy would help NASA secure its portion of the annual funding needed to operate its fleet of vehicles.[25]

Two years later, in 1972, new NASA administrator James Fletcher would officially inaugurate the space shuttle as the nation's flagship program in space. He cited four reasons the space shuttle was important and was the right next step in human spaceflight:

1. It was the only meaningful new human space program that could be accomplished on a modest budget.
2. It was needed to make space operations less complex and less costly.
3. It was needed to do useful things.
4. It would encourage greater international participation in spaceflight.[26]

Points 1 and 2 were never realized, because each flight cost nearly half a billion dollars. In fact, it was more expensive in actual dollars to launch the space shuttle than the Saturn V to the moon. The shuttles, which eventually flew 135 times, beginning with the launch of *Columbia* on April 12, 1981, turned out to be a much-reduced system compared to what Paine, Mueller, and just about every other expert had envisioned.

Mueller, in particular, had aggressively championed a fully reusable, multifaceted vehicle. His original design had an orbiter vehicle the size of an Airbus A320 commercial airliner equipped with jet engines so it could fly under its own power instead of gliding to the ground after reentry. A true spaceplane, it would have been mounted on the back of a booster the size of a Boeing 747 jumbo jet, which could also land on a spaceport runway under its own power. The critical selling point was that the planned rate of one flight a week would have amortized the cost of each mission low enough to justify its existence. Early projections were lofty by all accounts. Each vehicle was projected to fly into space one hundred times before being retired.

By these projections, the space shuttle could have been the first step in Paine's lofty plan to settle the moon and send human beings to Mars—the real goal behind his plan. In essence, it would have been a space taxi that ferried passengers and cargo between the ground and a space station in LEO. The station would have been assembled in orbit using shuttles as construction trucks. Astronauts with Buck Rogers-style portable jet packs (manned maneuvering units) would have served as construction workers. Once finished, the space station would have been an on-orbit Grand Central Station for launching missions to the moon, and later Mars, using the NERVA nuclear-powered rocket.[27]

The shuttles that eventually flew turned out to be marvels of aerospace engineering that did truly wonderful, useful, and unique things in space. They simply could not meet all of the far-reaching expectations and lofty goals that NASA and the aerospace industry had envisioned, however. With its design and development mired in delays caused by the late funding (coupled with Skylab's orbital decay, reentry, and demise in 1979), the shuttle had no place to go when it finally started flying. After just a handful of flights, decision-makers inside and outside of the space agency realized that it would never come close to achieving its operational cost objectives. With no space station in sight, the shuttles became the world's most amazing flying machines, without a convincing purpose. By 1990, Paine was one of the program's most vocal critics. When Vice President Dan Quayle asked him what the country should do about the space shuttle, he told him in no uncertain terms that NASA should phase it out "as quickly as possible."[28]

Besides the space shuttle and a space station, the third leg that would have completed Paine's triad to explore the inner solar system was human spaceflight beyond the moon, specifically, to Mars. Vice President Agnew was among the few in Washington who had gone on record to say that America, having gone to the moon, should now proceed to Mars. A Mars mission based on von Braun's research when he was at the Marshall Space Flight Center actually seemed doable for the 1979 launch window. But even if the technology were there, public sentiment was not. The lack of any real support on the part of the public prevented any action from taking place in fulfillment of the "Now on to Mars" charge.

In the months following Apollo 11, Paine began to come to the same disappointing realization. The public's reaction—a mere month after Armstrong and Aldrin walked on the moon—told him there was little chance that Congress would fund a Mars program any time in the near future. Furthermore, Nixon wanted nothing to do with Mars. Unlike the Kennedy and early Johnson years, there was no longer any deference toward the space agency on the part of the White House. NASA's fiscal year 1970 budget had, in fact, declined by 15 percent from the previous year. In 1971, it fell by another 15 percent. Any hope for a Mars mission died before it had a chance to be seriously debated on Capitol Hill.[29]

Congress clearly urged Paine to avoid extremes as well. Even Representative George Miller, chairman of the House Committee on Science and Astronautics and a friend of NASA, cautioned him:

> I understand very well the enthusiasm of those who draw on the experience of 1961 and propose a manned landing on Mars as our next great national goal. . . . But I do not at this time wish to commit ourselves to a specific time period for setting sail for Mars. I believe that there are many tasks that can be accomplished that will ultimately provide that capability, but will be less costly and will be necessary in meeting short term objectives. . . . I think it highly probable that five, perhaps ten years from now we may decide that it would be in the national interest to begin a carefully planned program extending over several years to send men to Mars.[30]

Paine was exasperated by now. The STG had deferred a much-needed decision on Mars until 1980 at the earliest. Worse yet, the White House and the Congress wanted no part of it. While it was a subject of long-range studies, the consensus in Washington was that any commitment to such a plan would be premature. There were still just too many unknowns about a long-duration space voyage. Instead, the "slow-paced approach" forced him to emphasize programs that had the best chance for support on Capitol Hill: earth sciences, technology applications, robotic planetary missions. These smaller programs ended up sustaining the agency. They bought time for NASA as it worked through the plethora of problems plaguing the space shuttle. In his view, dialogue on future plans for America in space had taken on a most pedestrian view. It had become too unimaginative and downright bureaucratic; the STG was the worst thing that could have happened to NASA at a time when it was trying to hold on to the momentum of Apollo, he vented in an interview in the fall of 1969.

He compared the critics of Mars exploration to those who, 170 years before, opposed the Louisiana Purchase. Daniel Webster declared then that the American West was a "howling wilderness fit for nothing but savages." Politicians nestled comfortably in their cocoons on the East Coast predicted that the West was "impossible for the young nation to ever inhabit." The purchase of Alaska from Russia in 1867 was derided as wasted money and "Seward's Folly." Even the exploitation of electricity was chided by a frustrated Michael Faraday as being "worthless as a newborn babe."[31]

Every time work was done in the process of exploring space, he continued, people on Earth benefited in ways that could not have been predicted. He spoke of Louis Pasteur's experiments in microbiology that led to pasteurization, vaccines, and reduction of epidemics in the field of medicine. "A jet engine is developed because the Nazis are shooting down our B-17s. We did not know then that it would lead to a global network of jet transportation and one of California's biggest industries. When Marconi studied radio signals, he did not know it would lead to color television. . . . If you believe the result of the Apollo program was that we brought back 800 pounds of rock, you really missed the whole point."[32]

His best ally in the Mars campaign was Wernher von Braun, one of the space program's most persuasive and visionary supporters. A series of articles by von Braun and science writer Willy Ley, published in *Collier's* magazine in 1952, had inspired people from all walks of life, from Walt Disney to Tom Paine. They had ostensibly turned science fiction into science, and encouraged readers to contemplate the limitless potential of space travel beyond LEO. Von Braun had lobbied for decades that the United States should establish a long-term infrastructure in space. The US, he said, should have the capability to do things and build things in space so that human beings could conquer outer space. With astronauts on jet packs, space taxis, and a mega space complex circling Earth, the US could be in a position to lead an armada into the vast expanse of the inner solar system. It would be much like the great voyages of Christopher Columbus half a millennium earlier.

A human mission to Mars was now the natural extension of having gone to the moon. Paine, as the leader of the country's civilian space program, believed the nation made a paramount mistake by not seriously considering it on the heels of Apollo. To him, asking the country whether or not it should go to Mars was another way of asking whether or not America was still confident enough to accept the most demanding challenge that it could.

He defended NASA's right to ask the president to "take the offer" to the American people:

> We made the offer . . . to put together a program and devote a decade or so of our lives to this tremendously difficult field, if the nation wished us to do so. I do not feel bad about having . . . put the offer to the nation. In spite of the fact that they turned me down, I suspect that we asked the right question or made the right offer, but they may have made the wrong response. . . . When you look at Mars, you cannot help but realize that Mars is not going to be settled as a national enterprise. Indeed, that would be grossly unfair to mankind as a whole. Everybody will want to participate and I think it is up to us to provide the leadership.[33]

He found himself in a quandary. He had a master plan for America in space all laid out, but Richard Nixon, and now even some of his own supporters, would not buy into it. He had no doubt that there would be footsteps on the surface of the Red Planet by the end of the twentieth century. The only question for him was whose boot prints they were going to be.

11

I ACCEPT YOUR RESIGNATION

What better way to have a pliant NASA than to have
a Democrat sitting there exposed to his people.
—Hans Mark

The rain was finally letting up. It had been pouring all morning, at times so hard that he could not see the thirty-eight-story-tall Saturn V rocket on Pad 34A three miles away. The clouds were still thick and low and the sky gloomy, but at least the countdown had now resumed. He turned to Pat Nixon and said there were no more planned holds. It would not be long now; the moon rocket should be on its way in less than ten minutes. The First Lady turned and told the president.

The low level of public interest in Apollo 12, America's second manned moon landing mission, did not concern Paine too much. Even the fact that lightning struck the rocket twice in the first minute of flight as it pierced the heavy clouds of a south Florida thunderstorm on its way to the moon had barely made the news. The *New York Times* reported in a sidebar that the launch generated what could only be called "scant enthusiasm." Network television broadcasts of the launch drew only a small, languid audience in the middle of the workday.[1]

Just four months earlier, *Eagle* had landed on the Sea of Tranquility. The anticlimax of the next mission was perhaps predictable, even expected, given the intense national preoccupation with history's first moon landing. But it was an unmistakable harbinger of the things to come. The American public quickly moved away from the diversion of the moon missions, back to its preoccupation with Vietnam, social justice, and civil rights. An irony was that, this time, there were no longer reports of demonstrations and protests of the kind that had surrounded the first landing. Marchers and antiwar protestors showed little interest in NASA's second attempt to land men on the moon. The sense in the country was that the Moon Race was over. America should now relax and retract, not only in space but around the globe.

Tom Paine was not about to relax. The year 1969 was a critical one for the space agency. The primary goal of the nation's first decade in space had been achieved with the Apollo 11 moon landing. Post-Apollo planning had already been well underway for over a year before Apollo 11 and 12 had even left the launch pad. Budget retrenchment was now accelerating throughout all branches of the federal government, as the new Republican White House sought to curb inflation through fiscal belt-tightening. Overseas, the cost of the Vietnam War continued to escalate with no end in sight. At home, the focus was on social priorities—poverty, a new war on pollution, and the state of the inner cities—not outer space.

Paine had recently announced plans for the agency to continue human missions to the moon through Apollo 20, as well as plans for several post-Apollo applications programs for the decade of the 1970s utilizing hardware developed for the moon landings. Proposals ranged from an Earth-orbiting laboratory like Skylab to a manned Venus flyby mission. These received a lukewarm reception on Capitol Hill. He now had to make some critical decisions. Having landed on the moon, NASA now faced major changes in program direction and staffing. What was most important to him was that these changes did not erode or undo the core competencies that the agency had built during the previous decade. Keeping intact the people of NASA, their knowledge, and their experience base were more important than the hardware that was being launched into space.

> I think that Thomas Alva Edison would have been a wonderful addition to our space program. We have many "Edisonian" types scattered throughout the space program at the present time, and the kind of inventive genius that he represents we have in NASA. . . . Anybody who can watch Apollo 12 come down and land within

> a thousand yards of a preselected landing vehicle that we put on the Moon a couple of years ago [Surveyor 3] and not be impressed about what this says about the reliability of American guidance equipment and the accuracy with which we can put payloads where we say we will put them obviously does not understand things.[2]

But to attract and keep the best people, NASA had to have another new program to work on. By the end of 1969, Paine was not at all sure what that program was going to be.

He was finding that President Nixon's approach to the space budget was mainly political, yet quite pragmatic—political regarding what the agency could do for him and his administration, and pragmatic in emphasizing fiscal responsibility and a balanced budget. The fiscal year 1970 budget was a very critical one for NASA. The agency needed enough funding in this first post-Apollo budget to get started on the right track for the 1970s. Paine was never able to get this point across to the White House, however. Nixon tended to treat space and defense together in his budgeting, and he had already decided that both would be cut severely in 1970.

Paine could already see that the agency was starting to lose the momentum it had gained from the success of the early Apollo missions. However, there was simply no national support for another major space program. With the war in Vietnam approaching a decade and the prospect of double-digit inflation looming, the mood of the country was one of impatience and uneasiness The political right was calling for more spending on defense even as the left called for increasing social spending. Nixon was at least candid with him about the budget for the space program. He viewed NASA as a means by which he could advance the popularity of his administration. While this had worked to perfection with Apollo 11, times were now different, the president told him.[3]

The nation's lukewarm response to "men in space" after the first few Apollo flights was due in part to the absence of a viable public response by the Soviet Union. The USSR, which had always answered the US, was now suddenly very quiet. The trouble was that the Soviets had not done anything dramatic in space for a long time. Nixon believed that when they did, it would change the national mood and reinvigorate support for the American space program. The inconvenient truth was that a new Soviet spectacular—an internationally stunning space achievement of the *Sputnik* and Gagarin kind—was what was needed to get America moving in space again. Until then, space (and defense) was going to suffer.[4]

This stance from the White House came about because Richard Nixon saw other nations as partners rather than competitors. Early in 1970, he asked Paine whether NASA could go as far as to accommodate foreign astronauts on US space missions. Paine quickly ruled this out as impossible for the remaining Apollo flights to the moon. He did shrewdly propose to DuBridge and the White House that if the president were to approve a second Skylab orbital workshop for 1975 or 1976, there might be a chance that a foreigner could fly on it. He never heard back from Nixon on that. The president did not take the bait.[5]

Richard Nixon was particularly sensitive to any action that could be construed by his political adversaries as taking money away from social programs. Paine, exasperated by the dangerously low funding for his agency, told Nixon's staff on several occasions that he hoped there might be less talk from the White House of the space program having to bear the brunt of cuts as compared to other federal programs in the next budget go-around. The White House had put him in this position, and it was one in which he had no room to maneuver. The White House and the Bureau of the Budget had already told him that it was in the agency's best interest to take all of the cuts in fiscal year 1971 and not postpone them into the future. The message was clear. Nixon felt that given the prevailing mood of the country, he was simply not going to put forth a daring new challenge in space.[6]

Paine had no choice but to take all the cuts immediately after Apollo 11 and 12. Spreading them out over several years was no longer an option. It would have bled the agency even more.[7] "The feeling I get . . . is that most of the senators, most of the representatives, have the feeling that they kind of like to vote for us, but they just don't feel the public support is there at home for the program."[8] He believed that most Americans would in fact support the space program if only they knew what they were supporting. But convincing the public that NASA was valuable to them had become increasingly difficult. Now, an entire generation openly questioned the American establishment. They directed their vitriol and distrust squarely against the government. Embroiled in the conflict in Southeast Asia, America was just not very enthusiastic about anything aerospace- or defense-related.

National party politics also encumbered him. In the early years of the space program, the Democrats had carried NASA. James Webb was very successfully able, by galvanizing strong Democratic support with the help of some Republican support, to move the agency forward. That was before the reversal brought on by the social justice movement and Johnson's Great Society fundamentally reshaped the nation. By the late 1960s, the social and political landscape of the country had flipped entirely. Now, he found that most of NASA's support—not counting the home districts of the agency's field centers—was coming from the Republicans, with help from moderate Democrats. Most ironic to him in the midst of this was the behavior of Ted Kennedy. The senior senator from Massachusetts was now NASA's biggest detractor, while his late brother, John Kennedy, had been one of the program's original champions.[9]

He was also in the position of having to deal mostly with Richard Nixon's inner circle of advisers—Peter Flanigan, H. R. Haldeman, and John Ehrlichman—rather than Nixon himself. (As history has since unveiled, this situation was by no means unique to Tom Paine during Richard Nixon's time in office.) Being a Democratic holdover from the previous administration did not help. He quickly found out the hard way how tightly Nixon's staff operated and how closely they guarded the president. As the agency's link to the president, he had to keep the lines of communication with the White House as open as possible, even if it meant that he had to deal with Nixon's famously recalcitrant staff—the gatekeepers of the Oval Office—instead of directly with him. This hurt NASA a great deal. Paine could never be sure just what was actually reaching the president's desk.[10]

With two moon landings now completed, Paine turned more of his attention to planning for the future. To assist him, he wanted a strong, large voice in Washington, someone who could immediately upgrade and sell NASA's strategic plans to Congress and the Nixon White House. He needed help, and Wernher von Braun fit the bill. The fifty-seven-year-old rocket scientist had fulfilled his life's work of building the rocket that sent the first human beings to the moon and was now seeking his next big challenge. He was unequaled among the NASA center directors, a larger-than-life figure even in an establishment now populated with world-renowned luminaries. Paine asked von Braun to join him in Washington. This he did, transferring from the Marshall Space Flight Center in Alabama on March 1, 1970.

"We needed to get a first-class, long-range vision up here so that, as we went down the road of the Space Shuttle and as we answered the question of what was man's future on the moon, what we were going to do in space stations, when were we going to Mars, we would have a person of real, proven vision," said Paine. Bringing his friend to Washington did that. In his new position, von Braun could now dedicate himself full-time to advising Paine on how best to position the entire agency so as to garner support for a Mars program; a manned Mars mission was both his and Tom Paine's top priority. Moving von Braun to Washington also broke up the clout that "the Germans" had built in Huntsville dating back to 1950—even before the formation of NASA. Paine gave him a lot of power at NASA Headquarters. The Offices of Space Science and Applications, Advanced Research and Technology, and Manned Space Flight all had contributing members on von Braun's management council.[11]

But bringing von Braun to Washington also garnered its share of criticism. An advantage of having him in Washington was his name recognition. Paine welcomed it as a chance to bring a strong personality to Headquarters. By now, he was finding it increasingly difficult to keep programs moving forward after Apollo 11, and wanted to increase the clout of NASA Headquarters in Washington. But industry observers saw the move as an attempt by Paine to return NASA to what critics called the "old arsenal concept" that Webb had been so successful at—growing government bureaucracy to increase funding. This was not his intention. He simply wanted to make sure that NASA Headquarters had the upper hand over the field centers in post-Apollo planning.[12]

He also called Neil Armstrong in Houston. On July 1, 1970, Paine appointed him as the agency's new deputy associate administrator for aeronautics. But Armstrong was brought in for a different reason. It removed him from active astronaut status. The first man to walk on the moon had wanted to fly again after Apollo 11 but realized that what had happened to John Glenn was happening to him. After Glenn became the first American to orbit Earth in 1962, NASA considered him "too valuable" to risk sending into space again—a veiled decision (purportedly from Kennedy) that he was not initially aware of.[13]

"I certainly had not said that I would not fly again. I was available," Armstrong recalled without much delay when asked about it in 2011.[14] Paine would stick to the official line, describing his position on the matter after the decision had been made:

> I looked around the agency to see where Neil might have the most interesting job. It seemed to me that that whole area of the new development of flying machines was the thing that might grab him. It seemed a natural. He was ready to leave Houston. He had finished in Houston and Washington seemed to be a good place. We also considered the possibility of putting him out at Edwards [Flight Research Center] which I think he would have liked too, although I do not think Jan [Armstrong] would have liked it so well. It is pretty remote out there.[15]

Armstrong oversaw several aviation research projects, notably the seminal development of a fly-by-wire control system (digital computer control instead of analog and mechanical control) for general aviation which was later perfected and adapted for use on the space shuttle. But the man who had the blood of a test pilot in him was now stuck "flying a desk" in Washington. He recalled that, from his vantage point, Paine showed no particular interest in what he had to say about the direction of aeronautics for the space agency. The two, in fact, had no significant program management interaction with each other after he was reassigned.[16] Just a year later, Armstrong resigned from NASA and returned home to teach aerospace engineering at the University of Cincinnati.

An important move involved longtime Associate Administrator George Mueller. On December 10, Mueller resigned and returned to private industry (the General Dynamics Corporation). He had overseen Project Gemini from its inception and was the one person directly responsible at the agency's headquarters for all human spaceflight programs. Webb had given him full responsibility for human spaceflight when he handpicked him to build up the Office of Manned Space Flight in 1963. But the Apollo 1 fire had put a rift between the two. Their thorny accord became infamous inside the hallways of the agency. Webb then made it a point to bring someone in from the outside who had no prior involvement with the program to be his deputy and bypassed Mueller. When Paine arrived, he had made it a point to try and develop a good rapport with Mueller.

It was clear to him that he needed Mueller's technical experience and well-known force of personality to keep Project Apollo moving.[17] When Mueller left NASA, space insiders speculated that a rift must have come between him and Paine. The larger disagreement was likely between Mueller and von Braun—who had Paine's full backing. Mueller believed that NASA's next step should be the space shuttle; von Braun, like Paine, wanted to go on to Mars. Paine was not disappointed to see Mueller go, but publicly credited him, nonetheless, with "having the creative leadership that enabled NASA to achieve the national goal set by Kennedy." In a way, Mueller prevailed. Within a year, with absolutely no sign that Nixon would ever sign off on a human mission to Mars, Paine had little choice but to endorse the plans for the space shuttle that Mueller had laid out.[18]

On April 11, 1970, Apollo 13 rose majestically from Cape Kennedy into the bright Florida sky. On board were Jim Lovell, Fred Haise, and Jack Swigert.[19] Their destination was the steep mountainous highlands of Fra Mauro on the Western Hemisphere of the moon. From Firing Room 1, Tom Paine, Vice President Agnew, and West German Chancellor Willy Brandt watched the picture-perfect liftoff through the panoramic glass wall of the Launch Control Center. Afterward, he took the microphone and introduced Agnew. The vice president congratulated the roomful of controllers on successfully sending three more American astronauts to the moon.

None knew then what was awaiting them. Two days later, Apollo 13 would turn into the most perilous flight of the entire lunar program. Four decades later, it is still very well known. "Houston, we have a problem" (although this was not the actual quote) has since become part of the vernacular—an unintended staple in popular culture made famous by social media and Hollywood. But the life-and-death drama, coming just nine months after the epic success of Apollo 11 and five months after the less heralded but equally successful Apollo 12, was very real. It was a sobering reminder that human beings still had a long way to go to conquer the perils of space travel.

Paine had just arrived back at home. He had stayed late to monitor the flight with other managers at NASA Headquarters after returning from the Cape. Still listening to the Mission Control voice circuit coming over the two "squawk box" speakers that the agency had installed in his house, he heard Swigert utter the famous line. Lovell quickly repeated it. There was a serious problem onboard the spacecraft. Fifty-five hours into the mission, one of the two oxygen tanks in the service module had exploded (later determined to have been caused by electrical arcing of damaged wires inside the tank). The powerful blast also gravely damaged the other oxygen tank. Lovell, Haise, and Swigert were running out of oxygen. Since the service module fuel cells used oxygen and hydrogen to produce electricity and water, they soon had neither.

At 2 a.m. EST, some four hours later, Paine took off in *NASA One* (a twin-engine turboprop airplane for official NASA use) in a driving rainstorm from National Airport. The pilot flew him straight to Ellington Field in Houston. Arriving at MSC, he was briefed on the situation. Bob Gilruth gave him a detailed account of what had transpired over the last eight hours. William Bergen, president of North American Rockwell's Space Division, and L. J. Evans, president of Grumman Aerospace Corporation, were also there. For the next four days, the team of contractors "froze" their support at three times the level that normally supported a mission. With the world, Congress, and the White House watching, the weight of the agency and the future of the space program would rest squarely on the performance of his people in the next seventy-two hours.[20]

With a moon landing now out of the question, the mission turned to one of crew survival. The guidance and control software had to be reprogrammed; timelines for consumables (oxygen, water, electricity) had to be reworked in real time. The lunar module had to support by itself a crew of three for three and a half days (it was designed for two people for two days). Engineers and astronauts exercised the simulators at Houston, Cape Kennedy, Downey, California, and Bethpage, New York around the clock. They had to simulate the post-explosion configuration of the Apollo spacecraft to come up with the lowest-risk way to bring the crew home. With the CSM completely powered down, Lovell, Haise, and Swigert relied on the LM as their lifeboat to bring them home. Mission Control created contingency checklists "on the fly," in real time. They were verified and reverified to make sure nothing had been overlooked.[21]

While the condition that occurred on Apollo 13 had been discussed in technical meetings, it was considered to be so remote that it was never simulated during training. There were a host of uncertainties. Houston could not even be sure that the heat shield covering the back of the command module had not been damaged in the explosion. After three days plus course correction burns using the lunar module descent stage engine, the spacecraft swung around the moon and made its way back to Earth and its fiery reentry into the atmosphere. As the spent service module was jettisoned, the crew saw that the explosion three days before had torn open an entire quadrant of the spacecraft. Tension remained high to the end. The communications blackout during reentry went half a minute longer than what the controllers had expected. But a pinpoint splashdown in the South Pacific eighty-eight hours after the explosion finally had Mission Control breathing again.

Nixon had called him at Mission Control in Houston and told him that he wanted to invite the family of the crew out to Hawaii upon their return. He also told Paine to go easy on the debriefings and to give the three plenty of time to rest and relax with their families. Paine had little desire to see the president turn the occasion into a public relations trip. The next day, the contingent boarded *Air Force One* at Ellington Field in Houston and flew to Hickam Air Force Base on Oahu to meet Lovell, Haise, and Swigert as they flew in from Pago Pago, American Samoa. With cameras flashing, Nixon presented each astronaut with the Presidential Medal of Freedom. He also declared April 19, 1970, a National Day of Prayer and Thanksgiving.

Paine was relieved. Nixon appeared to him to at least relish the moment. The president looked convincing as he expressed his strongest support yet for a vigorous, ongoing human spaceflight program. When Paine finally returned home to Washington a week after the accident, Barbara noticed that he had visibly lost more than a few pounds. (Tom Paine used to keep daily track of his weight on a graph. One can tell from the sharp dip on the graph when Apollo 13 happened.[22])

But he had to accept the reality that a moon mission had actually failed on his watch. The backlash was coming, and it was going to be directed at him. Before leaving Houston, he had directed George Low to put together an Investigation Review Board. He also ordered Dale D. Myers, who had taken over for George Mueller, to delay his return to Washington and remain in Houston and conduct his own independent study. What he specifically wanted from Myers was a recommendation regarding what to do about Apollo 14.

One week after splashdown, on Friday, April 24, he went before the Senate Committee on Aeronautics and Space Sciences. Apollo Program Director Rocco Petrone, Flight Director Glynn Lunney, Jim Lovell, and Jack Swigert followed him into the chamber. (The flight surgeon had requested that Fred Haise be excused from making the trip to Washington. He had developed a severe infection while on the mission.) Paine summarized the mission in two sentences: "The Apollo 13 mission was a failure. We did not succeed in America's third lunar landing attempt." Although the mission had failed, he endorsed before Congress the already emerging, increasingly widely held view that the recovery action following the explosion was a partial vindication. The fact that people had come through in the most stressing of situations turned out to be a "remarkable testimony" to the NASA team. He concluded his testimony by asking that Congress would at least not overlook that.[23]

Following a two-month investigation, on June 15, 1970, the agency released the findings of the Apollo 13 Review Board. He had already previewed the report twice. Before sending it to Capitol Hill, he had called the director of each NASA field center, telling them not to jump to conclusions. He also issued an agency-wide directive for managers at every level to disseminate the report as widely as possible throughout the aerospace industry.

He personally signed a copy for Keldysh and sent it to Moscow; perhaps the lessons learned might help prevent a similar occurrence in the Soviet space program. To the US Congress, he stressed that the lessons from Apollo 13 could be applied to disaster recovery in general. Complex engineering endeavors will always invite accidents. The most important long-term significance of Apollo 13, he testified, was in the lessons learned.[24]

What troubled him about Apollo 13 was the systemic failure inside his agency that had allowed the accident to happen. Despite all the rigorous management checks and quality control procedures, a hazardous condition had made its way 200,000 miles into space. It had nearly killed three astronauts. Privately, he told Low that he did not blame any one person or group for the failed mission. It would have been pointless. He had earlier told Congress that Apollo 13 was an unwanted case study on the inherent vulnerability of a large government organization to stumble. By itself, the agency could not have put a man into space, let alone land on the moon. Unlike the precise mathematics that govern the orbit mechanics of space travel, no equation could dictate whether NASA would succeed or fail. Rather, NASA's fate rested in the hands of a 300,000-person workforce. A breakdown in any phase of design, manufacturing, testing, or flight operations made the agency vulnerable.

Apollo 13 had exposed a weakness of NASA. The agency had to reevaluate the way it operated. He asked publicly whether it could reliably foresee, detect, and regroup to correct the deficiencies in this process. The *Washington Star* got it right with the headline "Paine Sees Setback to Program."[25] Just how much of a setback, he had no idea. He was still talking to the White House, but the dialogue became one-sided overnight. The damage was done. The ill-fated flight of Apollo 13 was the last manned mission with him as the head of the country's space program.

Tuesday, July 28, 1970, was much like any other hot, humid summer day in the nation's capital. It was also the day Paine announced that he was resigning as the third administrator of NASA. The Friday before, just as people were leaving their offices for the weekend, he had gone down the hall and asked George Low and Homer Newell to stay. They went to his office. He closed the door, sat down, and told them that he would be leaving the agency to go back to private industry. Having given no forewarning, he asked that they keep the discussion confidential until it was announced.[26] Over the weekend, he wrote his letter of resignation:

> Dear Mr. President:
>
> Please accept my resignation as Administrator of the National Aeronautics and Space Administration effective 15 September 1970. Now is an appropriate time for a change of command at NASA, and this coincides with my wish to return to private life.
>
> During my direction Americans orbited the Moon and walked on its surface, achieving our boldest national goal on time and within budget. We have made the transition to the post-Apollo internationally oriented space program of the 1970s, and the Congress has approved the new direction and pace in the 1971 budget. We will shortly publish a prospectus for man's conquest of space through the year 2000 which charts a long-range plan for future progress.
>
> The world can well be proud of the NASA team's incredible space achievements accomplished under four Presidents of the United States in twelve short years. Now the nation should press on boldly with the exploration of the universe as well as with the solution of man's problems here on the blue planet.
>
> It has been a privilege and honor to have led the nation's space program through critical times under two presidents. You have shown me every courtesy and consideration, as have your staff and the Congress. I am most grateful to you for having given me this unique opportunity to serve my country during mankind's first journey to another world.
>
> Respectfully yours,
>
> Tom Paine

Two days later, he called Willis Shapley, Julian Scheer, Clare Farley (special assistant to the administrator), and his secretary Betty Covert to his office. It was a Sunday. They suspected that something important was up. After he broke the news to them, he asked a stunned Scheer to review the letter and asked Covert to type it up. He also asked Covert to call Dwight L. Chapin, the deputy assistant to the president, requesting an appointment to meet with Nixon as soon as possible. Nixon, who was in San Clemente at the time, replied that Paine should fly out to California. The next day, he flew to Los Angeles with Julian Scheer. The following morning, they drove up the hill and met Nixon at the Western White House, where the president was vacationing. In a private meeting that Paine later described as short and cordial, the president accepted his resignation.[27] That afternoon, White House Press Secretary Ron Ziegler called a press conference and announced the news. Nixon accepted his resignation with the following letter:

> Dear Tom:
>
> I deeply regret that you will be leaving the government, but I accept your resignation as Administrator of NASA effective September 15, as you have requested.
>
> You have earned the gratitude of every one of your fellow citizens many times over for the outstanding leadership you have given to the nation's space programs. Your contribution to man's knowledge of the Earth as well as the heavens has been major, and the course you have done so much to set will help guide our efforts for years to come. The respect and affection of the colleagues and associates you leave behind will accompany you wherever you go, and I hope you will always take pride in your splendid achievements in behalf of every American and, indeed, in behalf of all mankind.
>
> You have earned a unique and permanent place of honor in the history of man's exploration. It has been a privilege to know you,

and to work with you, and to share with you the sense of excitement, adventure, and achievement that has marked this time of triumph in the nation's space program.

With warm personal regards,

Sincerely,

Richard Nixon[28]

Coming with no forewarning, his resignation shocked the agency. Even those closest to him had not expected it. Wernher von Braun professed that he was "totally surprised" when he heard the news. Directors of the NASA field centers expressed similar disbelief. "[He] bugged out," recalls Hans Mark, at the time director of the Ames Research Center. "At least [he] should have told us why he was leaving. He did not; he just left."[29] The immediate reaction among space insiders was that only a major disappointment of some sort could have made him walk away from leading the nation's space program, which was still near its zenith. But there were undeniable signs. Several major cuts had indeed been made recently, cuts that took away much of the program's momentum. The White House, while still publicly endorsing the rest of the ongoing Apollo missions to the moon, was privately quite worried about another high-profile accident. Paine's influence had been fading fast.

Nixon blamed him for Apollo 13. The president could not have been less enthusiastic about the grand post-Apollo plans that Paine was pitching around Washington. The canceled NERVA nuclear rocket engine program and the much-reduced space shuttle program had been major blows to the space transportation capability that Tom Paine had been counting on. NASA's proposal for a large-scale space station in Earth orbit also rang hollow. Nixon had already moved on from the tangent of space exploration, back to finding a way to extricate the US from involvement in Vietnam, assuage the civil rights demonstrators, and most importantly, win the upcoming presidential election.

He clearly left NASA before he was ready. Running NASA had become less rewarding in a time of draconian fiscal belt-tightening and disappointing public disinterest. To him, the space agency had taken more than its fair share of cutbacks. It had clearly responded to the request from the administration to cut its budget in a shift away from space to social issues. Nixon's primary interest in NASA was the prestige it brought to his administration, and while Paine had no problems with that, the president's commitment to a balanced and robust space program had to be called into question as reductions kept mounting despite his personal assurances to the contrary. Nixon had told Paine that America would continue to have a bold and balanced presence in space, yet did not point the country to any specific direction in space. Paine was thus left implementing a Nixon space policy with goals quite different from his own. There was more than a trace of irony to the whole situation. In the early days of the Apollo program he was never made to feel that the White House was looking over his shoulder or not supporting his decisions. But all that began to change as he tried to persuade the administration to start committing funds for the post-Apollo era. There was resistance on every level. The White House was simply no longer interested in what Tom Paine had to say.

He had come to NASA just before the social justice platform that defined the 1960s became an all-consuming fire that permanently changed the country. Although the administrator of NASA was, at least in theory, one of the least political jobs in Washington, he was a Democratic holdover who had served under President Johnson. After Nixon was elected, Paine had to transition the agency between two very different White Houses. He did this, and kept it on track for a moon landing. Having done that, he had fulfilled one of his own chief reasons for being at NASA.[30]

Despite never saying so openly, he was clearly bedeviled by the budget process. The continual burden of having to justify and defend new propositions in an ascetic environment took its toll.[31] As the *Chicago Tribune* wrote, he was "both disturbed and optimistic about America's future in space."[32] Paine was an able administrator, but he was at his best as a visionary. He was able to discharge the responsibility that landed a man on the moon, but could not push through a strong American presence in space after Apollo.

The moon landing had been an unequivocal success. Following the return of Apollo 11, he turned his attention to a broad yet specific set of goals for the decade of the 1970s. Congress took a small step forward by accepting the fiscal year 1971 NASA budget. While it lacked a sustained, major flagship program, NASA essentially received all the money it was likely to receive under the austere fiscal circumstances. In fact, Congress had appropriated 99 and 98 percent, respectively, of his budget requests for 1969 and 1970. After taking office, Nixon was able to exercise much tighter control of the budget. He reorganized the Bureau of the Budget into the Office of Management and Budget (OMB). This did not help the agency. Paine's frustration mounted as his priorities became impossible to reconcile with those of the White House. Strangling resistance by the OMB eventually left NASA with no flexibility for new plans or programs—a frustration that is still felt at the agency today.[33]

After Apollo 13 failed to land on the moon, Paine's voice with the White House waned quickly and considerably. "We were so damn suspicious of each other," said Paine candidly two weeks after tendering his resignation. More at ease now with his decision, he continued. "We in NASA knew that we could not leak any attacks on the president to the press, and we knew that we were being watched very closely. We had to be in a 'Caesar's wife' position; we darn well had to cooperate and we darn well had to demonstrate our loyalty to the president if we were going to be able to have the influence we needed to get the new program laid out and accepted by the White House."[34]

In retrospect, the most uncertain time for him in Washington was early on in the Apollo program. He had just arrived at NASA. The embers of the Apollo 1 fire were still warm. The spacecraft was being redesigned from the inside out, the lunar module was overweight, and the Saturn V rocket was progressing with known difficulties. Just two months later, Apollo 6, the second unmanned, all-up test flight of the entire Saturn V rocket stack, had experienced all kinds of problems. At that point, the setbacks could easily have derailed the program:

> There were some awful tough days. It really raised questions whether we knew what we were doing. When you had that many different failures in that many different areas, the question of

> whether or not we ought to fly a man on Apollo 8 was very much left open. In fact, we refused to take a position at The Cape, and the reporters all took the position that "you nuts would not dare fly a man on the next [Saturn] V mission." We said we cannot answer that, but would dig in and see what happened. So I remembered one of the first things as Deputy that I tackled, outside of getting to know the whole agency and trying to get the overall feel, was to dig into that Apollo 6 and satisfy myself as to whether or not these guys knew what they were doing. . . . You could not solve them all by always taking the conservative view. I think if we had not sent Apollo 8 around the moon, we could have had the Apollo 13 thing on the moon landing [Apollo 11].[35]

Tuesday, September 15, 1970, was the last day he walked the halls of NASA Headquarters as the administrator. As on his first day on the job thirty-one months earlier, he found a stack of papers waiting for him on his desk. He went through them and signed off on a few actions and contracts. Picking up the phone, he spent the rest of the morning talking to managers and colleagues from around the country. A final signature before leaving the building approved the prime contractor down-selection of either Bendix or Boeing to build the "moon buggy."[36] (Boeing would end up winning the competition over Bendix. The company would make four flight-ready lunar roving vehicles, three of which were driven by astronauts on the moon on Apollo 15, 16, and 17.)

Walking by Betty Covert, he asked if she had sent his farewell letter to all members of the House and the Senate. Over the weekend, he had also written letters of appreciation to General Antonio Perez-Marin of Spain, academician Mikhail Dmitrievich Millionshchikov of the Soviet Academy of Sciences, Henry Kissinger, and Under Secretary for Political Affairs U. Alexis Johnson. (It had been Johnson had who suggested the commemoration plaque and a goodwill recording be left on the surface of the moon on Apollo 11.) He then went over to Capitol Hill and, starting with Clinton Anderson, personally thanked each member of the Senate Committee on Aeronautical and Space Sciences and the House Committee on Science and Astronautics.

That evening, he and Barbara went to a dinner hosted by George and Mary Ruth Low at the Bolling Air Force Base Officers' Club. Overlooking the banks of the Potomac River, he eloquently addressed a large gathering of professional and personal friends from around the country, thanking them for their hard work and dedication to the nation's space program. It was not a poignant farewell. He admitted not knowing whether or not he was going to miss government life and Washington, but said he was sure that he could not have done anything better with his life during the previous two and a half years. The ideal time to quit a job never comes, he said, but this was the best time to do it. It was clear to the guests that evening that he did not want to leave NASA.[37] Finishing his speech, he looked up from the podium, smiled, and said, "Looking back at the things we have done, there is little to be said; perhaps wait and let history be the judge of decisions that have been made."[38]

Four more Apollo flights landed on the moon after he left NASA. He spoke through his writings, speeches, and interviews, saying that the epochal achievement of human beings journeying away from the planet was about more than just the breakthroughs in engineering and science. Those were impressive, for sure, and would be remembered long after the strange hardware in museums had awed the curious of the coming generations. But twentieth-century history would record that Americans, working together purposefully in a time of peace, had opened the age of human exploration of the solar system. Aware that it was probably too early to comprehend the full meaning of the achievement, he proposed that the same history would inevitably draw one of two conclusions from the moon landings. One would attribute to them the Columbus type of legacy, as a trail blazed on a new ocean that others then followed and pushed forward. The other would compare them to the spectacular but one-time conquest of Mount Everest.[39] Would the NASA astronaut of the twentieth century be remembered as Christopher Columbus or as Sir Edmund Hillary? Now that America showed that it could go to the moon, would it still aspire to? Paine never saw the moon as a celestial Mount Everest. For the Apollo program to realize its full legacy, it had to be the Columbus type of undertaking, he reflected. Earth's moon was but a stepping-stone to the planets. Interplanetary travel was next; Apollo was just the beginning. The fact that America had met such a challenge in the past by no means portended that the country would choose to do so in the future.[40]

He once pointed out to NASA historian Gene Emme that the United States landed on the moon in less than ten years under the same civil service regulations that ran the Post Office. Why had the country embraced Kennedy's lunar challenge? It may have been that the United States believed that such a success would give the nation greater world recognition than the Soviet Union. It mostly did. He also believed that the United States, as a world superpower, needed to undertake great endeavors and give itself the great challenge to lead humanity outward in a new age of discovery. Marco Polo, Magellan, Balboa—all would have applauded President Kennedy's bold challenge, he believed.[41]

In 1961, Kennedy was able to do the kind of convincing that usually took a great war or a national crisis to be effective. "In the sense that the space program mobilizes the energies of hundreds of thousands of skilled people and gives direction to their work," said Paine, "it is a kind of warfare without loss of life, a kind of war effort that adds enormously to our wealth."[42] "In many ways, the most interesting aspect of the Apollo program is that you [had] 300,000 people who worked together for a number of years and worked together in a way that truth and honesty and technical rigor were absolutely essential to the success of the program. . . . But any place where we allowed any lack of rigor to enter in, as in the Apollo 13 . . . incident, we were going to get found out and the program would not succeed."[43]

To him, the nation's goal in the 1960s to put a man on the moon by the end of the decade was the right objective at the right time. Had a bolder objective like a human Mars mission been tried, the cost and difficulty would have probably been too much. And had a less ambitious course been taken, the US would not have been in as strong a position. Beyond reaching that goal, the real, long-term payoff of the moon landings was that the United States learned how to develop the enabling technologies that could one day be used to move human beings farther into the solar system.[44]

He gave three reasons as to why the Apollo program worked. First, people knew what NASA was trying to do, and it was widely accepted as a national commitment, even by the ordinary "man on the street." Second, bold yet feasible goals were set and the end dates publicly announced. Finally, the progress toward these goals was dramatically visible to all. There was no way for the agency to back out unless it admitted failure. As a well-accepted, long-term national commitment, Project Apollo was thus able to command adequate resources and attract the best talent of that generation. Once it got started, a challenging deadline maintained the pace. This forced tough decisions. And most importantly, milestones were visible to an entire nation, if not the world, as a measure of progress.[45]

In the autumn of 1970, he left behind that challenge, the awesome responsibility that went with it, the weight of the space program, with its triumphs and setbacks, and the burden of captaining that ship known as NASA. For now, they were in his rearview mirror. For a little while, at least, he could relax a bit and take a break from all of that.

12

A LITTLE BETTER FOOTING

You know the great respect and affection that I will always have for General Electric and its great people.
—Tom Paine to Reginald Jones, CEO of General Electric, in 1976

"I must say that I would . . . be willing and even anxious to return to Washington at some time if there were some need at the national level for my talents. The job at the moment is maybe to get myself on a little better financial footing." He was talking to Robert Sherrod, the longtime *Time Life* journalist who was in New York to interview him. A year had passed since Tom Paine left NASA and the tangos inside the Washington Beltway. From behind his antique mahogany desk, he looked out the window at the sprawling skyline, the concrete and steel high rises, and the taxi-filled streets of Midtown Manhattan thirty-eight stories below. Thirty Rockefeller Plaza was a long way from Federal Office Building Six. The landmark Art Deco skyscraper that was now his office was home to the Radio Corporation of America and General Electric. Leaning back in his chair, he had no problem telling Sherrod that it was "a lush job."[1]

Months earlier, he and Barbara had moved to East 56th Street, just a few city blocks away. General Electric, his old company, had wanted him back. In the summer of 1970, he was still deeply immersed in working the return-to-flight issues following the Apollo 13 failure when he received a surprising, unsolicited job offer from the Xerox Corporation. It was a very generous offer from the Fortune 100 document technology products company. He did not want to just say yes or no, but it got him thinking. Greta was in college, and George, Judith, and Frank were all in private school. "It makes you feel good inside, . . . reassured that you are still respected."

He picked up the phone and called Jack S. Parker, the vice chairman of the board of General Electric and an old friend. Was there anything at GE that he might perhaps consider before responding to Xerox? As a matter of fact, there was, Parker told him. He quickly put Paine in touch with Fred J. Borch, GE's chairman of the board. Borch, on the other end of the line, offered him the opportunity to head up all of the company's worldwide electric power generation businesses.

While still a premier global corporation, General Electric in 1970 was in the midst of an unprecedented period of instability, mergers, and divestitures. One big problem it was struggling with was the acute shortage of top management people. There was, in particular, almost no one in Paine's age bracket with his high profile, experience, and qualifications. The global high-tech conglomerate also had recently made some highly publicized and openly unprofitable moves in its nuclear power business, moves that were roundly blasted by critics in the industry. The worst thing was that Westinghouse and Babcock & Wilcox, its chief rivals, had won some major legacy contracts away from GE. The company needed to make a change.

Borch's offer appealed to him on several key levels: it was in the field of engineering and technology, dealt with aspects of energy production, and had nothing to do with outer space. "One of the strictures I put myself in was that I would work outside the aerospace area," he said. "I did not want to be put in the position of talking business with the people I once worked with."[2]

He told Borch that it might be best if he waited until at least early the following year so he could continue to try and make the case with the White House to support his post-Apollo plans. With Apollo 14 scheduled for January of 1971, that would also allow him to still be at NASA when Project Apollo returned to flight. But Borch needed someone to step in that summer. Hubert W. Gouldthorpe, the vice president and group executive of power generation, was stepping down. Paine finally told Borch yes, but to give him three more months. He would leave NASA and make himself available by the first of October.

General Electric Power Generation was the industry's standard-bearer in the world of electricity. As the largest power generation business in the world, it employed some 38,000 people on five continents. With worldwide manufacturing and licensing agreements, GE had produced, in the years since World War II, $1.5 billion worth of power generation equipment. Its long history of working with turbines had also made it into a world leader in the manufacture of jet engines.

GE presented Paine with a big challenge. The energy crisis had been brewing for years. Now it was replacing Vietnam in the national headlines. Circumstances would take an abrupt turn for the worse with the 1973 OPEC oil embargo that would force the price of imported crude oil to triple virtually overnight. This affected nearly every sector of life, from manufacturing to food processing, housing, and of course, transportation. Long lines of cars unlike any that had been seen before appeared at gas stations across the country. The automobile industry changed for good. More fuel-efficient compact vehicles replaced the large luxury sedans that came out of Detroit. Environmental activism, too, gained momentum, and had become quite vociferous by the early 1970s. Efforts to develop clean air and clean energy drove nearly every facet of engineering, research, development, and manufacturing.

Paine quickly settled into his new job. In 1971 and through 1972, he came up with a strategic plan that he believed would help confront the energy crisis head-on (and increase GE sales in doing so). It began with expanding the technical diversity of General Electric in a time of national need. He put more company dollars into research and development in the advanced technologies of coal gasification and hydrogen fuels. Properly developed, they could provide an environmentally friendly way for the US to decrease its reliance on fossil fuels. He wrote to the White House and recommended that the Nixon administration open up new domestic and offshore drilling. A key component of the plan, he told Nixon, was to construct the Alaskan oil pipeline (now called the Trans-Alaska Pipeline System).[3]

His other objective was to expand GE's nuclear power market even in the midst of the heated opposition by antinuclear activists and scrutiny from the political left. He spoke to the industry regularly about the need for the country to increase its dependence on nuclear power based on its clean track record and reliable scientific principles. The best way was to start cutting down on federal government procedures, rules, and regulations. Once that happened, more power plants should be put on the country's power grid as quickly as possible. Critics publicly challenged him in the New York press and business publications. But he maintained that nuclear power was the way to go. It was by far the most proven and affordable form of clean energy available.[4]

On a trip to Santa Barbara, he met with Philip Luttenberger, the new manager of TEMPO. He had an important task in mind for his old group. Could they, he asked Luttenberger, come up with a working plan to develop an integrated electric power and water supply for the entire North American continent that, when implemented, would be viable to the year 2000? If successful, such a plan would give the company a huge advantage over the other power companies, making it possible for GE to grasp the market and help guide the country out of the energy quagmire using the commercial sector.

TEMPO met his challenge. Luttenberger's group gave him a comprehensive plan that was very detailed and, he believed, marketable. It proposed ways that General Electric Power, working with its suppliers, could aggressively increase research and development on new sources of energy such that they might be ready for use by the early 1980s. An incremental buildup would continue the growth to meet the sharp rise in consumer demand that was expected through the end of the twentieth century. Accompanying this would be changes in federal government regulations concerning the development and mining of new resources. The timing of the plan was perfect. It was ready to take to the White House, and he proposed it to the new Federal Energy Administration (later the Department of Energy) that President Nixon created in May 1974.[5]

In the spring of 1973, he directed GE to put more money directly into the research and development of synthetic fuels. By his estimation, "synfuels" held many answers for the energy crisis. Prospects for the artificial fuel sources, he said, were so promising they could extend "across the entire non-electrified energy spectrum, from home heating and the clean production of steel and fertilizer to auto and jet fuels."[6] This was not a trade secret. It was a well-established position throughout the industry that the first to achieve a breakthrough and make synfuels profitable would hit a "gold mine" in returns. The company was cash strapped, however; the board of General Electric did not agree with him and were much more tepid in their enthusiasm. But a potential breakthrough would have a multitrillion-dollar impact on global power generation. He maintained that it was the right thing to do in a time of national crisis, and kept the investment programs moving.[7]

Energy was the top domestic concern of the country in the early 1970s. Economic stagflation had hit the United States hard, and the same was happening in other parts of the industrialized world. The US was the greatest consumer of energy in the world at the time. Scientists projected that with population growth, annual energy requirements would more than double by the year 2000.

President Nixon was by then mired in the Watergate scandal; criminal investigations had already been underway for the better part of a year. But the president still had to address the issue of energy. On June 29, 1973, Nixon created the Office of Energy Policy. Its first priority was to implement a national policy to tackle the energy crisis. A few weeks earlier, Paine had received a call from Peter Flanigan. The White House wanted to know if he was interested in the job as the nation's first Energy Czar.[8] They discussed the matter on subsequent occasions, and Flanigan wrote to him, saying that the job would undoubtedly "both please and displease pretty much every faction of America." Surprising no one, Paine turned it down. He thought that it would be the most political job in all of Washington. (The appointment eventually went to John Love, the moderate Republican governor of Colorado, who was very popular in the western states for championing the protection of the environment.)

It was still early in 1973 when Paine finally landed in Moscow; it was his first time in the Soviet Union. For this important company occasion, General Electric needed someone with name recognition who had more than just a cursory experience dealing with the Russians. When Reginald Jones called on him, he said yes.[9] He was happy to go. (As he later recalled in a lighter moment, it finally gave him a chance to use the bit of Russian that he had learned years ago at Stanford.)

On January 12, 1973, on behalf of GE, he signed a landmark agreement with the USSR Council of Ministers' State Committee for Science and Technology.[10] The industrial accord brought the two adversaries together for one of the first large-scale, global cooperative efforts in the midst of détente. The synergy would go a long way to jointly advance each nation's electric power generation capability. The timing of the accord was not accidental. It came at a time when Soviet-American relations were just starting to thaw. Just two weeks later, the Paris Peace Accords would end official US involvement in Vietnam. The Soviets wanted the accord badly. They gave Paine and his delegation a very warm reception and dinner after the ceremony.

He was spearheading a historic agreement. It allowed the two adversaries to exchange specialists and equipment and to share the results of their research. For the first time since before World War II, the State Department allowed the doors to be opened for a large-scale commercial venture. Soviet engineers could visit GE facilities across the US and Americans could visit power stations in the Soviet Union. The pooling of talent helped both sides. A solution was needed to meet the anticipated demands for much higher power consumption for both sides. Besides providing a working basis for exchange of technical data, he hoped that the agreement would quicken the pace of research and development in both countries while broadening its reach. The cooperation benefited both sides, since they did things differently. The Soviets tended to prototype at the earliest date possible, often disregarding cost. The Americans, on the other hand, preferred to try experimental designs until a more economical solution was obtained.[11]

After Paine returned to the US, the White House and State Department claimed victory. It was front page news. President Nixon praised the GE initiative from the Rose Garden, even as newspaper editorials blasted the company's motives, on top of accusations of corporate greed.[12]

GE was also a major player in the Far East. The potential returns from the Asian market were especially lucrative for the Power Generation Group. Paine worked very closely with Hoyt P. Steele, the vice president and general manager of the GE International Sales Division, to develop a strategic marketing plan. Together, after months of negotiations in 1971 and 1972, they secured a long-term manufacturing and service agreement with the Taiwan Power Company. Taipower, as it is known throughout East Asia, was a pioneer. It went from zero capability in the years after World War II to become one of the leading indigenous electrical power companies in the world. In that time, modern electrical power totally transformed the Chinese island republic off the coast of mainland China. By 1970, it had become a leading global exporter of everyday consumer goods—everything from tennis shoes to portable radio sets.[13]

The negotiations delighted Paine. The landing on the moon had just captured the imagination of the island's people. It had garnered him a reputation as the mysterious phantom behind the success of Apollo. From his days in the Pacific, he had had an affinity for Asian culture and its unique elegance. This sense only deepened after he worked with Republic of China Vice President Yen Chia Kan. On a personal level, he became deeply fascinated with the wisdom of the traditional Confucian analects. When Yen found out, he had them translated and sent to Paine. Among his favorite was a saying that he perhaps thought described himself. He wrote it down in illuminating cursive: "Is he not a man of complete virtue who feels no discomposure, though men may take no note of him?"[14]

One business area that he grew was GE's renowned industry training program. For decades, the company had sponsored international training, inviting engineers from other countries to America, and in exchange, sending its own engineers overseas. Skilled people came to train and work at locations across the country. Locally recruited and trained by GE, workers became proficient electrical technicians, construction supervisors, and site managers. Puerto Rico had hosted the first program. The US territory was, at the time, struggling to convert third-world agricultural laborers into skilled workers. The higher-technology industrial tasks needed an infusion of new talent. The program steadily grew from there. Under Paine, the Power Generation Group greatly expanded the program. Citizens from many nations, including the Republic of China, Japan, the Philippines, Honduras, and Mexico, came to the United States for training. It was a good business development move and one that he promoted enthusiastically with yearly trips abroad to each of the partner countries.[15]

Five months after his breakthrough with the Soviet Union, General Electric promoted him to senior vice president for technical planning and development. It made him the fourth-ranking executive in the company. The title meant that GE now looked to him to understand and predict the effect of near- and long-term developments in the rapidly changing global world of high-tech. The much broader role affected all GE business groups, from aircraft engines to nuclear reactors to home appliances and even Christmas lights.

Chairman and CEO Reginald Jones had given him the position based mainly on the strength of his ability to do long-range planning. In his new job, he had to do one thing very well: forecast the market and recommend a direction for the company.

The situation at GE in the mid-1970s was not unlike that of many large enterprises trying to survive in a time of global economic recession. Investment was low, inflation was high, and the unemployment rate was approaching 10 percent. Because of short-term pressures on financial returns, attention at the top of the company had for years been directed mainly at just the next few fiscal quarters. A comprehensive, long-range strategy for the company as a whole was not very well defined—nor was it greatly desired, as he found out. He told Jones of his surprising find, that there was "a near consensus among our executives that a corporate strategic plan for a diverse company like GE is neither possible nor desirable."[16]

Why was GE struggling? He came up with not one but a handful of reasons. Foremost among them was the fast-increasing prominence of liberal social reform ideology and environmental activism that had begun a decade earlier. There was also a new scrutiny of the government; the Watergate scandal was a catalyst but it was not the only one. The national media had turned on large businesses with stories of greed and corruption. Lastly, Communist bloc nations were entering the free world trade market for the first time. All of these adversarial factors turned the social and political climate of the country against big businesses.[17]

The situation was even tougher on GE because it had heavy ties to Washington and the defense lobby. Paine believed that the company had been too slow to adapt to the changing times and not selective enough in how it used its resources. Available capital was becoming more and more limited. It was impossible for the company to grow. Later that year, he met with Jones and the board at a retreat at the company's new headquarters in New Canaan, Connecticut. Jones wanted to focus on the long-term growth of GE. But Paine was frank with him. Before that could happen, GE must "batten down the hatches" and learn, as NASA did, to operate in a transparent, very public fishbowl.

The following spring, Paine led a complete, top-down review of the strategic plan and the entire management structure of the company. The exhaustive analysis took up the better part of 1975. It was a challenging task. He directed the other executives to take a long, over-the-horizon look at the technology trends and sociopolitical climate confronting the company. Could GE start to adapt so it would be in a better position to grow going into the 1980s? And if so, would it? He gambled and was boldly direct with the CEO, asking Jones, "How will we manage GE to achieve?"[18] He had put himself in an unenviable position. In January 1976, he had a sit-down with Jones and the Board of General Electric and pointed out to them what he believed were long-standing cultural shortfalls and problems with the company, along with his recommendations.

Then he resigned and left the company.

Tom Paine wanted to be a chief executive. After twenty-five years at General Electric, Jones was not going to give him that opportunity. He did not make Jones's short list of handpicked candidates in the CEO's carefully crafted and complex succession plan.[19] The previous fall, he had put his name in with several executive personnel consulting firms. He had put down in writing exactly what he wanted: an opportunity to be a "CEO of a major enterprise requiring technological, operating, and international leadership, and in which high-level business and government experience would be valuable."[20]

On November 22, 1975, he received a call from the executive selection firm of Heidrick and Struggles. They wanted to know if he would like to meet with the management board of the Northrop Corporation in southern California.[21] Since it was right before Thanksgiving, he told them he would fly to Los Angeles after the holiday. On Monday, December 1, he met with the company's senior executives at the California Club in downtown Los Angeles. They got straight to the point: Northrop wanted Paine as their president and chief operating officer.

With 26,000 employees and $2 billion in annual sales, Northrop was one of the largest defense contractors in the US.[22] Founded by aeronautical engineer Jack Northrop in 1939, the aerospace powerhouse was famous for its line of legendary warbirds dating back to World War II. In 1944, it had introduced the P-61 Black Widow, the first radar-equipped night fighter to enter service in the Pacific. Other aircraft followed, including the first all-weather strategic interceptor, the F-89 Scorpion, squadrons of which the Air Force stationed in Alaska as the country's first line of defense against Soviet first strike nuclear-capable bombers. The revolutionary YB-49 Flying Wing would lead to the development of the B-2 Spirit strategic stealth bomber four decades later in the 1980s. And the T-38 Talon, still the most successful jet trainer of all time fifty-six years after its introduction.

The Northrop Corporation especially wanted his experience in the overseas market. The company had been trying to regain its reputation in the foreign sales establishment after the troubling events of 1974. That year, the fallout from the Watergate scandal had revealed that company president Thomas V. Jones had made an illegal contribution to the reelection campaign of Richard Nixon. The wrongdoing had made some news as part of the Watergate coverage, but the subsequent congressional investigation made headlines. A much more egregious impropriety was revealed. It showed that the company had paid some $30 million in bribes to government officials from Indonesia, Iran, and Saudi Arabia in an effort to secure foreign sales of military hardware. A class action lawsuit was filed against the company. The court ruling forced Jones to step down as president of Northrop. It also stipulated that a new president be named by July 1976.

On February 18, 1976, the board named Paine the president and chief operating officer of the Northrop Corporation. He had no connection to the scandal, and the board of directors brought him in to repair the damage.

The company was trying to establish its F-20 Tigershark program, and was having great difficulty doing so. The fighter aircraft was designed by the company to compete exclusively on the foreign export market. Based on Northrop's venerable F-5 fighter that first entered service in 1962, it was equipped with the latest in electronics, radar, weapons systems, and engine technology. But to make the aircraft, various parts of the company had to come together and operate more fluidly. Paine had to ensure that they operated seamlessly as a single entity.

Every division and subsidiary of the corporation reported directly to him. This provided him a panoramic view of the state of the company. He began to identify areas of weakness and strength and those where he thought strides still had to be made. The key was to have one operating model that could be implemented across all divisions of Northrop. This was central to his plan to unify the company. As each division executed its operating plan, he made sure that it was aligned with the central corporate plan.[23]

He created a five-year schedule with quarterly milestones that could sustain Northrop in the midst of the government's draconian cuts in aerospace and defense. Under President Jimmy Carter, the period from 1977 through 1980 was extremely lean for defense contractors. Northrop was one of the hardest hit in the industry, along with Southern California in general. Using his plan, by 1980 the company would be in a position to pursue the new, large procurement contracts that were expected to come out of the Defense Department under a new administration in Washington.

But on March 1, 1982, Paine stepped down and left Northrop. He believed that he had fulfilled his obligation to the board of directors that had hired him and the six-year contract he had signed. Under him, the company had more than turned the corner. Most of the weaknesses that he had identified had been rectified. Reforms central to the common operating model that he had developed had been implemented at various levels throughout the company. And most notably, Northrop had defeated its chief competitor, Lockheed, to win the highly coveted prime contract to build the B-2 Spirit stealth bomber—still the most advanced and only operational, low radar visibility strategic bomber in the world today. The company's growth was unprecedented. *Dun's Business Month* would name it one of the five best-managed companies in America.

As he sat gazing out the window at the busy highway to Santa Monica, not far from downtown Los Angeles, the box that he had been waiting for finally arrived. It had only been a couple of weeks since he stepped down from running the Northrop Corporation. He was still setting up his new office when the package from the printer came in. It was the ten boxes of stationary he had ordered a few weeks back. Opening the box, he saw the letterhead that he'd designed neatly printed on heavy sheets of letter paper:

> Thomas Paine Associates
> *High Technology Enterprises*

The small embossed, well-known picture of Buzz Aldrin's boot print on the moon at the top of the letterhead looked quite attractive. It was a nice touch, he thought. He sat down, pulled out one of the sheets and began writing. He now had ample time to do what he relished most: chart a way to the stars.

13

PIONEERING THE SPACE FRONTIER

Our progress in space, taking giant steps for all mankind, is a tribute to American teamwork and excellence. . . . We can be proud to say: we are first; we are the best; and we are so because we're free.
—Ronald Reagan, State of the Union Address, January 25, 1984

The space program had sunk to a new nadir. It had reached a point where it could not just keep coasting. Space exploration was not a priority for Jimmy Carter. Tom Paine knew that; others knew it as well. He knew that the policies of the thirty-ninth president of the United States were too weak to guide the nation's space activities on any realistic level. Even as an interested private party on the other coast of the country, he did not accept the speeches coming out of Washington that argued for American leadership in space at the same time that programs were being slashed one by one. He wanted back in the action.

Upon taking office in January 1977, the president's science advisor, Frank Press, had advised Carter that he should not commit to any specifics in space without first doing the research. But the problem with that was that the inherent lead times involved for large aerospace programs were by necessity measured in years and not months. The deliberately cautious approach stopped any possibility of real progress dead in its tracks. Congress, including a good number of ranking Democrats led by Olin Teague of Texas and Don Fuqua of Florida, still favored a goal-oriented space policy over one of general principles with no clearly defined programs. But the White House now wanted to cancel the space shuttle even before its first flight and then eliminate all "unnecessary" NASA funding.[1]

"American space policy [is] a leadership issue," said Paine during this time. He pointed to "a lack of national consensus and weak leadership from the White House [having] led to drift and indecision" on the direction of the US in outer space. "Ringing rhetoric proclaiming United States leadership in space is no substitute for plans and programs," he told the Congress on July 23, 1980. "In my view, it is self-delusive to give lip service to leadership while avoiding initiative and commitment."[2]

The nation's space program needed to go in a new direction. In his opinion, the "business-as-usual" way of doing things had to go. High-level committee planning, coordination, and management at NASA had also been notoriously ineffectual. The agency needed to be more ambitious, he said, but should also be disciplined enough not to propose future programs without the budget increases required to successfully execute them. He proposed that NASA start promoting a multi-decade plan as quickly as possible. If it came out with such a plan, it would put the onus on a new White House and new Congress to make a decision regarding space.[3]

NASA needed the help and welcomed his voice. Paine stayed busy and made himself heard. The agency's deputy administrator at the time, Hans Mark, recalled years later that "basically, Paine was an influence pretty much for whatever we wanted to do. He wrote a lot of op-ed pieces and a lot of articles in science[,] . . . all of which helped."[4]

On November 4, 1980, Ronald Reagan carried forty-four states and won the 1980 presidential election in a landslide. The former governor of California knew little about the space program, but he liked the NASA folks, endorsed the agency's mission, and demonstrated more interest in it than any president had since John Kennedy. National security was his top priority. He knew that a strong defense meant that the US had to do better in outer space, that highest of the high ground. It was crystal clear to everyone in his administration and throughout the capital that he came into office wanting to do all he could to defeat the "Evil Empire" (Soviet Union) and end the Cold War. It was, from day one, the main goal of his presidency.[5]

Paine didn't know Reagan (they had met only once in passing years before in California), but he liked what he was hearing from the new president. The unmistakable and refreshing policy shift coming from the new Republican White House was exactly the kind of leadership that he was hoping would make a difference. Members of national grassroots, pro-space groups also seized on the opportunity. They wasted little time and stepped up their call for the nation to revitalize its space program.

The list of advocates leading the charge was long. Joining the familiar names were younger lobbyists, enthusiasts, and newly elected officials. They allied with former Apollo astronauts, science writers, military officers, and top aerospace executives in a groundswell of activity designed to seize the moment. They came at the issue from different angles. All had their own vested interests in a renewed US space campaign. They had voiced their position during the presidential campaign and put the issue in the Republican Party platform. Soon after the election, they gained entrance to critical transition activities that were taking place at all levels in Washington.

George Low, who was leading the NASA transition team, gave him a call. Paine went straight to the point with his former deputy: "No move by the new president would more quickly and visibly reverse America's technical decline than his announcement of a bold, forward-looking U.S. space program." An Apollo-like flagship program would go a long way, he said, to restoring the US presence in space, win back international prestige, and lift national morale.[6]

Others, such as defense policy expert Karl G. Harr Jr., who had helped frame Eisenhower's response to *Sputnik*, personally encouraged Paine to take the case as strongly as possible to Washington. "The returns from space investments are among the most beneficial realized from any federal program. You know that and I know that, but it is not getting across to the public. We presently lack the solid constituency needed to support the type of program you outlined. I think we of the space advocate community have a big job to do in communicating the value of expanded space operations."[7]

The timing was important. The new administrator of NASA, James M. Beggs, felt that he could only count on having a four-year window to sell a program to Reagan. There was no guarantee that there would be a second term, nor any further interest if the president did not approve a plan during his first term.[8]

Paine stepped up his campaign. He recommended that NASA, with support from the White House, develop a comprehensive, long-range plan out to the year 2000. President Reagan could then announce it to a joint session of Congress at the end of the first space shuttle flight, which was scheduled for early in the spring of 1981. Implementing such a plan could lead to the settling of three worlds (Earth, Moon, and Mars) by the year 2050. The effect on national standing would be evolutionary, felt for a century. With the beginning of the shuttle era, he was convinced that Reagan had an outstanding opportunity before him to show leadership early in the area of space exploration.[9]

For the military, space was the focus of the weapon system that critics derided, with more than just a hint of sarcasm, as "Star Wars." Under the president's Strategic Defense Initiative (SDI), the US planned to construct a high-tech missile shield in space to render nuclear weapons impotent and obsolete. Using directed-energy weapons like lasers and high-powered microwave beams, the orbiting shield would shoot down Soviet ballistic missiles outside the atmosphere before they could deliver their nuclear payload. That was the focus. But on the civilian side, Paine saw an opportunity. He was counting on the fact that Reagan also wanted the country to build a strong national program to exploit outer space.

The pro-space community now had a renewed sense of solidarity. Reagan's people were just fine with that. They used the public lobbying to gain more support for SDI—the real objective driving the White House. Their view was not so much that it was great to support the space program, but that it would have been bad not to support it.[10]

Congress noticed. Led by Democratic Senator Ernest F. (Fritz) Hollings of South Carolina, it placed a provision in the space agency's 1985 authorization mandating that the president establish a special commission on space.[11] They wanted an answer as to what the US should do in space beyond building the space shuttle and a space station. What would be the outlook on space twenty years in the future? National leaders on both sides of the aisle wanted to see just what America could aspire to do in space into the twenty-first century if given the opportunity.

On Friday, March 29, 1985, the National Space Club did its part and presented President Reagan with the Robert H. Goddard Memorial Trophy. The timing of the award was deliberate. The nation's most prestigious award in spaceflight is bestowed to an individual or group of individuals who have accomplished something especially meritorious in the field of aerospace—LBJ was the only president, besides Reagan, who has been so recognized. Stepping up to the podium, Reagan accepted the honor to enthusiastic applause. Reagan then announced that he had selected Tom Paine to lead a National Commission on Space, or NCOS. In the words of the president, he wanted the commission to "devise an aggressive space agenda to carry America into the 21st century," using what he believed in—the incentives of individual freedom and profit to tame the frontier of space.[12] The White House was looking for the perfect civilian complement to SDI. Paine got what he wanted. His appointment by the White House was nothing if not predetermined. While he did not lobby for it, he had done enough to get the White House's attention.

"It was a consensus by NASA, my office [Office of Science and Technology Policy], NSC [National Security Council], everybody that was interested," recalls Reagan's science advisor, George A. (Jay) Keyworth II. "I don't think there was any controversy at all about Tom Paine. . . . NASA needed a new vision. NASA was incapable at that time of coming up with its own vision and it needed some outside stimulus. Tom was well respected. I knew that he had a view that did not believe that the space station was the end-all of the space program, and very strongly felt that we needed to think bigger."[13]

The clout of the nation's first-ever commission on outer space was due to its commissioners. Appointed with Paine were fourteen other American luminaries from the fields of science, engineering, and academia:

- Dr. Laurel L. Wilkening (Vice Chair)—Planetary scientist and Vice President for Research at the University of Arizona
- Dr. Luis W. Alvarez—Nobel laureate at the Lawrence Berkeley National Laboratory

- Neil Armstrong
- Dr. Paul J. Coleman—Assistant Director of the Los Alamos National Laboratory
- Dr. George B. Field—Senior Physicist at the Smithsonian Astrophysical Observatory
- Lieutenant General William H. Fitch, USMC (Ret.)—Former Deputy Chief of Staff of Marine Corps Aviation
- Dr. Charles M. Herzfeld—Former Director of the Advanced Research Projects Agency
- Dr. Jack L. Kerrebrock—Former Associate Dean of Engineering at MIT
- Ambassador Jeane J. Kirkpatrick—Former US Ambassador to the United Nations
- Dr. Gerard K. O'Neill—Physicist at Princeton University
- General Bernard A. Schriever, USAF (Ret.)—Former Commander of the Air Force Systems Command
- Dr. Kathryn D. Sullivan—Space shuttle astronaut
- Dr. David C. Webb—Chairman of the National Coordinating Committee for Space
- Brigadier General Charles E. (Chuck) Yeager, USAF (Ret.)—Test pilot

They were joined by nonvoting congressional advisers and ex officio members from the president's cabinet. A staff of sixteen staffers and writers supervised by Executive Director Marcia S. Smith from the Congressional Research Service rounded out the commission.[14]

They wasted little time. The two-day kickoff meeting in May, held at the National Air and Space Museum, focused on how the commission would address issues relating to direct NASA activities, non-NASA civilian activities, and national security planning.[15] By the end of the meeting, the commission had a good idea of its initial strategy. They refined it over the next month. By the second week of June, they had developed a detailed outline. It was specific enough that Paine next wanted to get some feedback from the administration to see if they were on the right track. On June 10 over lunch, he showed it to Richard G. (Dick) Johnson from the OSTP. Johnson's reaction was that it was "fully consistent with OSTP directions and initiatives." He was, however, not yet convinced that Johnson's boss, Jay Keyworth, would feel the same way. Over breakfast the next morning with Wilkening and Smith at the Washington Hyatt Regency, they penned a plan to expand upon the outline so as to get a draft to the White House as soon as possible.[16]

By the middle of July, the initial working draft was ready for review. Paine took it to Jay Keyworth. He specifically wanted to know from him whether or not some of the initial concepts that the commission came up with would be too bold for the president. If so, the NCOS would need time to make the necessary course corrections so that its recommendations would be acceptable to the president. They went over the initial findings in Keyworth's office. Keyworth liked it. At one point, he astounded Paine by assuring him that nothing could be too bold. This emboldened Paine. He left the meeting confident that they were on the right track. He wanted to push the envelope now and recommend an extremely aggressive space program, one not just for the next twenty years but for the next half-century.[17]

From the outset, he steered the group to consider Mars as the goal. This was not a major debate, as it was a fairly obvious objective to everyone on the commission. Other than the moon, Mars was the only place suitable for human colonization. Just as important, it was the one place that still stirred people's imaginations. Near-Earth asteroids that only astronomers had ever heard of, or even the moon itself, just could not elicit the kind of romanticism that Mars did. "If you believe in people in space," said Laurel Wilkening, vice chair of the NCOS, "then Mars is where you are going to end up."[18]

This was backed up by fifteen public forums that the commission held across the country. The first of these was in Los Angeles on September 13. The hearings then moved around the country to other locales—Boulder, Boston, Iowa City, Seattle, Ann Arbor, and lastly, Honolulu the following January. They were town hall–type forums that gave voice to the ordinary citizen. Schoolteachers, especially, turned out in great numbers and took part in the debate. Feedback from the public was surprising, better than expected. It showed that the average person on the street could still be captivated by the romance of human spaceflight. The American public wanted more than just the handful of shuttle flights a year that NASA was sending into space. From the commission's research, Paine was convinced that the age of Apollo could be revitalized over the horizon of the approaching new millennium. What he wanted most was for people across the country to start identifying with the space program and think about outer space again.[19]

The other strong theme that came out of the public forums was a desire for commercial spaceflight in the form of space tourism. While not tourism per se, NASA already had a program to fly noncareer astronauts in space. This was the original intent of the payload specialist position on the space shuttle. They were meant to be everyday professionals like scientists, journalists, or teachers who would train with NASA to fly once or twice on the shuttle. The public made it clear that they wanted more, much more. Paine also wanted to see ordinary "people with a minimum amount of training from all walks of life" in space. No other way could better open up outer space to the general population, he maintained. Nothing else would get people talking about outer space faster. He said with an unmistakable grin on national television that he would personally welcome the chance to fly on the shuttle if given the opportunity.[20]

As they discussed how best to get their message across, he urged his fellow commissioners to refocus on the endgame of what they were trying to accomplish. Paine wanted them to think about the best way to make their case and to put themselves in the position of those whom they were trying to reach. The commissioners also had to come to an agreement among themselves on the specifics of the major themes. "As with any group of people, there were many different options on every item of discussion," recalled Armstrong, on the challenges facing the commission in reaching a consensus.[21]

At first, the staffers did most of the research and writing. The group met, usually at an office in L'Enfant Plaza in Washington, every few weekends to review their progress. As the report began to take shape, something just did not seem right. The opinion among the commissioners was that what was being put together on paper was not really communicating the full intent of the commission. While the technical points were good and well thought-through, the way the case was being made was just "not living up to the demands of the audience of one," namely, the president.[22]

Jeane Kirkpatrick, a very close friend of Ronald Reagan, was convinced that the report was not going to be forceful enough for the Oval Office. Paine had put her on the commission specifically because of her White House connection. From that point on, the commissioners changed their approach. "The commission took it over, and we were the authors, the whole commission," said Wilkening. "[Jeane] made a key directional change for us." By March 1986, the report was ready.[23]

Paine and Wilkening were ready to present it to the president on April 11, when the White House unexpectedly postponed the appointment. Three days later, Reagan ordered airstrikes on Libya (Operation El Dorado Canyon) in retaliation for the terrorist bombing of West Berlin earlier in the month. They had to wait three more months. On Tuesday, July 22, 1986, with new NASA Administrator James Fletcher and Acting Science Advisor to the President Dick Johnson enthusiastically looking on, they handed the president a copy of *Pioneering the Space Frontier: An Exciting Vision of Our Next Fifty Years in Space.*

The two-hundred-page book was the glossy end product of a year's worth of research, discussions, and writing. Besides the elegant prose, it had vivid, original artwork by the notable American painter Robert McCall and astronomy artist William Hartmann. The attractive presentation clearly bore the fingerprints of Paine's exploratory vision throughout. It described in surprisingly persuasive detail how a wide road to explore and settle the vast frontier of space could be paved by the United States in the latter half of the twentieth century and early twenty-first. Looking fifty years into the twenty-first century, *Pioneering the Space Frontier* tried to answer the question, what will 2035 be like? It was aspirational even by his own standards, he admitted right after the work came out.[24]

It was as difficult to project in 1985 what the year 2035 would be like as it was to predict, in 1935, how 1985 was going to be. Back then, Pan American Airways was just pioneering the transpacific route in the *China Clipper*; commercial air service was still in its experimental infancy over the North Atlantic. McCall and Hartmann's paintings complemented the report with vivid artwork that captured the forward-thinking spirit of the commission. It depicted the evolution of 1950s science fiction turning into the reality of space travel in the span of just a few decades.

Before they could make their recommendations, the commissioners had to first reach a consensus on several key points. All agreed that technology had advanced to the point where the human race could now move outward as a species destined to expand into a new and very different world. In that regard, the country's pioneering heritage made it fitting that the US should lead this movement. Space would be the new frontier of the twenty-first century just as the West was a century or two earlier.

They penned for twenty-first-century America a daring and sweeping "Declaration for Space," much in the same way that the Founding Fathers penned the United States Declaration of Independence two hundred years before:

> Lead the exploration and development of the space frontier, advancing science, technology, and enterprise, and building institutions and systems that make accessible vast new resources and support human settlements beyond Earth orbit, from the highlands of the Moon to the plains of Mars.[25]

The plan was a straight-to-the-point road map for how people could leave Earth and settle the inner solar system. What made it stand out was that it was not merely a generic road map; it was a customized map with program details and schedules. From seminal, small lunar colonies of twenty people in the year 2005 would come larger outposts. These pioneering colonists would set up pilot facilities to mine the moon for resources, manufacture spaceflight hardware by taking advantage of the low gravity, and conduct organic research for food, oxygen, and water. They would grow food in self-sustaining habitation modules. Rigid and inflatable biospheres would use completely self-contained artificial biological systems to recycle organic waste products. A variety of plants would metabolize carbon dioxide to generate oxygen. It would be the cleanest of clean environments using just indigenous resources. This would then be followed by regular journeys to and from Mars by the year 2015. Like the "Great Ladies," the majestic ocean liners of days past, tomorrow's spacecraft would have all the necessities for the nine-month voyage each way. By 2035, the first Martian of Earth ancestry would be born on the Red Planet.

The fantastic plan was notable because it was specific and because it had a timetable (not unlike Project Apollo). Most importantly, no large or immediate increases in space spending would have to take place. Instead, funding would rely on a steady growth in the space budget of 2.4 percent each year. This would match the historical annual increase of the gross national product of the United States in the latter half of the twentieth century. NASA's budget would only have to remain at a steady 1 percent of GNP. Paine had said since the days of Project Apollo that this would be enough to free up the agency to pursue new initiatives.

Pioneering the space frontier would start with a space station in the 1990s. It would then proceed outward to create a "Bridge between Worlds." This "Highway to Space" would rely on seven technologies that, in 1985, were either already available or in various stages of development: (1) the hypersonic aerospace plane; (2) advanced chemical propulsion rocket engines; (3) aero-braking for orbital transfer; (4) long-duration closed eco-systems; (5) ion electric propulsion; (6) nuclear power generation; and (7) space tethers and artificial gravity.

Twelve milestones would serve as signposts along the journey. Step-by-step, they would culminate in a Martian colony by the year 2035:

- Initial operating capability of a permanent space station in low-Earth orbit by the year 1995.
- Initial operating capability, by the year 2000, of a dramatically lower-cost transport vehicle (much cheaper than the space shuttle) to deliver cargo and passengers to and from LEO.
- Beginning in the early 1990s, additional modular transfer vehicles to move cargo and people from LEO to any destination in the inner solar system.
- Between the years 2000 and 2005, construction of a spaceport in LEO (modified from the space station *Freedom* that was then being planned) to serve as a gateway into the solar system.
- Operation of an initial lunar outpost by 2004 and pilot production of rocket propellant from lunar surface natural resources beginning in 2008.

- Initial operating capability of a nuclear-electric rocket for long-duration scouting missions to the outer planets after the year 2000.
- First shipment of nuclear shielding mass-harvested from the moon after 2005.
- Deployment of a spaceport in lunar orbit to support expanding human operations on the moon after 2010.
- By 2012, initial operating capability of an Earth-Mars transportation system for robotic precursor missions to Mars.
- First flight of a ferry spaceship to open continuing passenger transportation between Earth orbit and Mars orbit in 2015.
- Human exploration and extraterrestrial prospecting from astronaut outposts on the Martian moons Phobos and Deimos, and then on Mars itself in 2017.
- Initial operation of the first Martian resource development base to provide oxygen, water, food, construction materials, and rocket propellant in 2022.[26]

On this "Highway to Space," three new classes of spacecraft would replace the space shuttle. The commissioners maintained (correctly) that the biggest holdup to long-term space exploration was the state of space transportation in 1985. (The same holds true in 2017.) Low-cost, routine access to low-Earth orbit was the cornerstone on which the entire plan rested.[27] Since none of the commission's recommendations could be carried out by the extant space shuttle, they charged NASA to retire it as soon as possible, but certainly no later than the year 2000.[28]

The cost of going into space was the big problem—as it still is today. For the plan to work, cargo operations to orbit had to come down into the neighborhood of about $200 per pound. (The average cost to deliver one pound of payload on the space shuttle was over $10,000.) "The cost of moving goods into orbit has to be reduced to 1/10 of that of the shuttle," said Paine at the time, "then reduced again to 1/5 to 1/10 of that level."[29]

To take passengers into space, a new reusable launch vehicle, or hypersonic aerospace plane, would be needed. The commission proposed five years for its development. That would lead to a down-select on a vehicle design by the year 1992. Then a completely new orbital transfer vehicle would move people and cargo between low-Earth and high geosynchronous orbit.[30] Finally, a revolutionary, newly designed spacecraft would be used to journey between Earth and Mars. It would be the most technically advanced of all the spacecraft designs. The spacecraft would use nuclear propulsion mass drivers to escape Earth's gravitational field. Ion propulsion would then sustain it on its long-distance interplanetary voyage in deep space beyond the moon.

For this to happen, the space program had to have an evolutionary leap in heavy-lift launch capability. The commissioners proposed to leverage commercial space transportation for this purpose. An important conclusion was that the country must engage private industry to develop a robust commercial space transportation market sooner rather than later. In the 1930s, US airmail, using canvas-covered biplanes flown by barnstormers like the daring Jack Knight, not only showed that government contracts could get the mail delivered but also made possible the civilian airline industry. In the same way, NASA could use the strength of its existing infrastructure to demonstrate and fly the new technologies. It could then pass them to the commercial sector where their full potential could be realized.[31]

Paine believed his commission had come up with a good road map and the necessary signposts to guide the United States and international partners in the colonization of the inner solar system. But the big question was, could it be done? He explained why he believed it could from a metallurgical perspective:

> We don't think that there is anything on the Moon or Mars we don't have on Earth. But the difference is that the Moon and Mars are regoliths, where all of the elements are just sort of spread out into a surface powder. It may very well be possible that rather heavy elements such as iridium and platinum—which, through the processes of geologic change, have been taken down to the center of the Earth—may be quite abundant on the surface of the Moon.
>
> It isn't clear that the Moon will be good for mining, but all the metals you want are there. Since the Moon has no atmosphere, you don't have to separate iron oxide. You can just go along with a

> magnet or run a conveyor belt past a magnet, and metallic iron will jump right up. The Moon is ideal for powder metallurgy because its metals already are powder. The rocks of the Moon are composed of about 40% oxygen, and since oxygen is one of our spacecraft fuels, one of the first products to be exported from the Moon will be rocket fuel.
>
> Let's have private industries supply that oxygen to NASA on the Moon. In other words, NASA could give the contract to a university to develop a process; then NASA could put together a pilot plan. But once feasibility has been demonstrated, NASA should give a contract to Exxon or somebody to supply so many gallons per day of liquid oxygen on the Moon. . . . Natural resources here are at the bottom of an enormous gravity well. To use any of that material in space would cost you a hell of a lot to lift it up. If you want to use materials to build things in space for use in space, it makes much more sense to use materials you can pick up on the Moon or the asteroids.[32]

P*ioneering the Space Frontier* drew mostly guarded praise but also had its fair share of critics. *Aviation Week and Space Technology* called it "a valiant try." Opponents disparaged its predisposition to human missions at the expense of scientific exploration. The commission was widely applauded for its use of public opinion and the budding technology of the Internet. Even critics, to their credit, reluctantly praised the diversity of its research, and its visionary approach that merged science with science fiction. Proponents urged national leaders to give the report more than a cursory nod, lest it be reduced to the end product of a year-long "what-if" exercise.[33]

Most critics, however, called it too ambitious and too improbable. Their main concern, predictably, was with the estimated $700 billion expenditure that would be needed between 1995 and 2020. Amortized over that time, the cost came to some $27 billion per year, or about 40 percent more than NASA's budget actually was in the year 2016. Those critical of the commission's work thought, however, that that was too hefty a price tag to pay and would deprioritize more practical goals with more immediate returns. "I am worried this is going to do damage to the cause," said Thomas Donahue, chairman of the National Academy of Sciences' Space Science Board. "You can never justify this in terms of science."[34]

But it was never Paine's intention to propose the plan as a scientific endeavor. He was thinking in much grander, inspirational terms. Human society and civilization itself would be impacted immensely. Robotics, artificial intelligence, transportation, medicine—all would undoubtedly have to be revolutionized. New space-based industries with global reach would be created to lift the world economy. Unlike the Space Race of the 1960s, major international participation would now play an important and long-term role. The ultimate impact might not be seen for decades or perhaps even a century, but humankind would one day undoubtedly reap the benefits. It would be like the history-changing undertakings of the past across the vast expanse of the oceans, replayed now in outer space.

James Fletcher immediately seized on the work to help make his case for more near-term funding. He called it "the course for America in space." "The goals of the report are ambitious, yet achievable, but only if we maintain the momentum of the space program," he said. "We are confident that we will overcome our short-term hurdles and get the program back on track."[35]

What had been missing for a decade in Washington was any coherent, long-range planning in the area of space exploration. This had not been a priority of any sort since the days of JFK. And even then, the president used space not for its potential benefit to society but as a chessboard on which to play out the Cold War.

The National Aeronautics and Space Act of 1958 that created NASA had made provisions for a National Aeronautics and Space Council (NASC). The intent was to keep attention focused on space at the national level. But under presidents Johnson and Nixon, the council's influence waned. Nixon then eliminated the council in 1973. Civilian space policy in the Ford and Carter years was usually addressed at a low level by the OSTP. This improved somewhat under Reagan, who addressed civil space as part of national security. The emphasis was heavy on defense and light on exploration, however, and was still not an ideal setting for rich funding.[36]

Paine's commission recommended that the NASC be immediately reestablished, this time in a stronger form. To be truly effective, the council had to have more stakeholders. For the type of priority, commitment, and leadership required to make it happen, civilian space would have to be elevated to the same level as national defense (Department of Defense), national commerce (Department of Commerce), and urban development (HUD).

The following year, the US Congress introduced language in the 1987 NASA authorization bill to at least reestablish the space council; but Reagan vetoed it. Paine knew then that he had overestimated the president's interest. Ronald Reagan's interest lay elsewhere, with SDI, Gorbachev, and the economic recovery. He was the only one who could have put the country on a new path into space, but instead, paid little attention to it.

Paine liked Reagan. They were both spirited Californians, and the taming of the American West had been an inspiration to both. He even had *Pioneering the Space Frontier* made into a 30-minute video in which outer space was portrayed as the new West, narrated as an American saga by Louis L'Amour, Reagan's favorite author.[37] Like the unique setting of the American West of the 1800s, conquering outer space was painted in *Pioneering the Space Frontier* as challenging, inspiring, good, and profitable—a vast frontier with unlimited entrepreneurial potential. America could make the world a better place by leading humanity out on this new, high-tech adventure.

But none of it was quite persuasive enough. The White House thought that the recommendations were just too ambitious. They were not politically or fiscally tenable in 1986. The timing was also off. The final report was released only six months after the *Challenger* shuttle disaster had killed seven astronauts on their trip into orbit. That effectively took away any hope that *Pioneering the Space Frontier* would capture much attention in Washington. The commission had considered the timing, but had decided that while the launch disaster was certainly a staggering blow and a national tragedy, the nation would recover and that the fifty-year time span of the plan would still be relevant.[38]

Reagan's response to the commission was delayed, and given the post-accident quandary of *Challenger*, the report languished in the hands of the White House staffers and was quietly shelved. The fading outcome was not altogether surprising. Paine had proposed a monumental commitment of the nation with (in hindsight) little political backing. The result might have been different had he asked for less.

"I thought his views of the 'possible' were very optimistic," said Neil Armstrong later. "Having been involved in the ups and downs and cancellations of many programs, I expected that there was little likelihood that such a bold program would find sufficient political support for implementation. I believe he was very disappointed when *Pioneering the Space Frontier* was reduced to human operations limited to low-Earth orbit."[39]

14

CHIEF MARTIAN MONSTER

Tom got that in a big way. He was a space guy first,
but he was definitely an internationalist in his outlook.
He knew that it had to have that kind of high-level importance.
—Lou Friedman interview with the author

In the summer of 1985, NASA Administrator James Beggs asked Tom Paine to represent the space agency as the US endeavored, once again, to open talks with the Soviet Union. A resumption of dialogue with the Russians on outer space was by now long overdue.

After the joint Apollo-Soyuz mission in July of 1975, NASA had negotiated a series of incremental and what could only be called minimalistic agreements with the Soviets. They included continuing research into a common spaceflight rescue capability and the exchange of technical data in space sciences. President Richard Nixon (and later President Ford) showed some signs of commitment to a joint space program in the spirit of détente and directed NASA to formulate a space station agreement with the Soviets. For the next eighteen months, Arnold Frutkin made progress such that the Soviets were ready to negotiate the language of the agreement.[1]

But the US position changed entirely when Jimmy Carter took office in 1977. Almost immediately, the new administration told NASA to scrap the negotiations. There would be no dialogue for four years. The complexion of the relationship then changed again when the Reagan administration tasked NASA with reopening the talks. Ronald Reagan wanted to bring the Soviet Union to the negotiation table as part of the larger effort to end the Cold War.

That October, Paine flew to Moscow for a week-long summit. High-level space officials from both sides met for the first time in nearly eight years. With him were seven members of the US Congress, US Ambassador to the Soviet Union Arthur A. Hartman, and Apollo-Soyuz astronauts Tom Stafford and Deke Slayton.[2] It was a breakthrough for the United States—the first formal talks on space between the two superpowers since the days of Apollo-Soyuz. The discussions began promisingly, and were quite extensive in technical details. The delegation interacted with every level of the Soviet space establishment, from cosmonauts to members of the Academy of Sciences.

The US wanted to concentrate on cooperative opportunities for the decade of the 1990s, and steered the talks in that direction. Options that were on the table included a robotic Mars sample-return mission and an astronaut exchange program. Congressman Bill Nelson of Florida, who led the talks, proposed a high-profile joint mission for 1992. The timing would celebrate the thirty-fifth anniversary of *Sputnik*. It would also be the seventy-fifth anniversary of the Russian October Revolution and the 500th anniversary of Columbus's discovery of the Americas. A "Year of International Space Cooperation" could go a long way, he said, to showcase the progress of détente between the two superpowers.

Paine thought that the talks were going quite smoothly. Then Andrey Gromyko arrived. Sitting across the table from the Americans in the Kremlin's landmark Ekaterininsky (St. Ekaterina) Hall was the imposing president of the Presidium of the USSR Supreme Soviet and a dozen stoic-looking officials from the Communist Party Central Committee.

Gromyko had zero interest in sharing a space program with the United States. His real interest in any cooperation on outer space was to use it as a bargaining chip in the Soviets' diplomatic campaign against the SDI. If the space-based, high-tech missile defense system were to be deployed, it would clearly shift the balance of power to the United States.

The timing of the space summit was key. October 16 was just one month before arms reduction talks between President Reagan and General Secretary Gorbachev were set to convene in Geneva. Gromyko showed up to do one thing: influence the agenda beforehand. The summit quickly digressed from cooperation in outer space to angst over SDI. As the morning progressed, he grew more and more incensed, finally pounding the table and shouting indignantly: if the US continued to pursue what he viewed as the militarization of space, then "all opportunities for cooperation will be blocked!"[3]

The talks collapsed. Paine, Nelson, and the others flew home without any kind of agreement. But despite the conflicting agendas, Paine learned quite a bit about the Soviet space program itself that had previously been shrouded in mystery. For example, the Soviets had the same difficulty as NASA in making a spacesuit that minimized crew exertion during spacewalks. This had been a problem dating back to the first EVAs in the mid-1960s. Carrying out seemingly mundane tasks during spacewalks had brought several astronauts and cosmonauts dangerously close to the point of total physical exhaustion.

He also learned that the Soviets had upgraded the solar cells of their *Mir* space station from silicon to the more advanced gallium-arsenide. GaAs, as it is known, can absorb light more efficiently and resist radiation damage, making it ideal for space electronics applications. This greatly increased the efficiency of converting sunlight to electricity on Russian spacecraft. Unlike US spacecraft, which used batteries or fuel cells or a combination of the two, Soviet spacecraft since the 1960s had been powered by solar energy. Future space stations would depend almost entirely on solar energy for its power, and the Soviets were clearly far ahead of the US in that area.

The Soviet Union also revealed that it had shifted the focus of its planetary exploration away from Venus to Mars. Paine was troubled to see just how much progress they had been quietly making in the field of bioscience. The technology would be critical for long-duration human flights to Mars. Similar research had been nearly dormant at NASA for a decade. Their research into artificial gravity, closed ecology life support systems, water recycling, and synthetic protein manufacturing were all well ahead of the US.[4]

And finally, he saw firsthand what he and other experts had long suspected about the demise of the Soviet shuttle program. Their attempt to build a shuttle had ground to a halt following the economic turmoil of the mid-1980s. Only one vehicle was being built (and it looked suspiciously like the NASA shuttles, even down to the tile patterns). When it was finally done, *Buran* (meaning "snowstorm") ended up flying into space only once on an unmanned, automated test in November of 1988. The entire program was canceled shortly thereafter as the communist state collapsed abruptly, bringing the Soviet Union to an end. *Buran* and all its support equipment ended up in an abandoned hangar at Baikonur.[5]

When he arrived back in the US, Paine wrote down how disappointed he was over failing to reach an understanding with the Soviet leadership. He was now serving the nation in a very different capacity. As a private citizen campaigning for a stronger and more purposeful space program, he did not have to worry about political fallout. National leaders in Washington used him as an international goodwill ambassador anytime outer space was the topic.

The Soviet Union, in particular, had held him in high esteem since the days of the moon landings. In October 1987, they invited him back, this time to receive the prestigious Tsiolkovsky Space Exploration Medal acknowledging his lifetime contribution to the progress and history of human space exploration.

He found a changed Moscow this time, and a refreshing openness in the Soviet Union. There was a clear desire to seek international cooperation in space that was absent before. The effect of glasnost was unmistakable. This time, SDI was not vehemently denounced as militarization of space or a roadblock to peaceful cooperation. Not only was there no linkage, the topic of SDI never even came up.

The trip was particularly memorable, the sightseeing relaxing, and the people very hospitable. One afternoon, he and Barbara ducked into a large church in Leningrad to escape a fierce thunderstorm. They suddenly found themselves in the midst of a beautiful Russian Orthodox wedding. Sitting down quietly in the back, they were soon deeply entranced by the most heavenly setting of flickering altar candles, devout worshipers kneeling before antique icons, the angelically veiled bride in her radiant white gown, and the groom in his traditional morning suit carrying their candles down the aisle. Chanting priests in gold brocade robes read from a silver-chased Bible. There was stirring a cappella music, heavy golden crowns above the bride and groom's heads, emotional relatives and friends weeping tears of joy, and even a fainting bride. It struck a chord with him to think of the bond that ties together humanity on planet Earth.

The crowds in Moscow and Leningrad, he observed, were much more smartly dressed and in better spirit than what he remembered of the old Soviet Union. People were very friendly everywhere. A Red Army officer on their train from Leningrad to Moscow was so delighted with his *Amerikanski* compartment-mates that he insisted on fixing the pillows to make Barbara more comfortable. The babushkas in the dining car went out of their way to make sure that they got a good dinner.

They attended several memorable ballet performances (without any official escort) and dined at one of Moscow's new regional restaurants, the Baku. The cabaret that night featured excellent Azerbaijan cuisine, wine, and music. A bottle of chilled vodka instantly cured his bursitis, or so he claimed. Glasnost had also brought Moscow its first pizzeria. An authentic Chinese restaurant was also being planned. At the Pushkin Museum, they toured a Chagall exhibit. There, Muscovites enjoyed paintings far removed from the style of socialist realism—no more portraits of peasants with painted smiles joyfully brandishing sheaves of wheat.

Glasnost was transforming all parts of the Soviet Union, even its once highly secretive space program. The Academy of Sciences showed him their space agenda through the year 2000 and wanted to know if he had any recommendations. In a twist of irony, the Soviets had studied in detail the findings of the NCOS and *Pioneering the Space Frontier*. He found their long-term, step-by-step agenda for space exploration eerily similar to the plan his commission had outlined: develop a heavy-lift launch vehicle (the *Energia* rocket); deploy near-term probes to the moon and Mars; launch an armada of robotic missions throughout the inner solar system; and send human expeditions to Mars by the turn of the twenty-first century.

That the Soviet Union possessed a long-range plan only underscored the absence of one for the United States. The woes and murky future of the US space program could be put squarely on the shoulders of Washington, he said when he returned to the States. "There is nothing wrong with the aimless US civilian space program that decisive leadership and support won't cure. NASA's problem is confined to a tiny triangle bounded by the

White House, Capitol Hill, and the Pentagon. . . . All that the U.S. space program needs is decisive leadership and support. . . . I do not believe that officials within the Beltway understand the widespread disgust and anger of our space science community over the destructive mismanagement of NASA's once great exploration programs."[6]

On July 20, 1989, Vice President Dan Quayle, NASA Administrator Richard H. (Dick) Truly, and the crew of Apollo 11 gathered on the steps of the National Air and Space Museum behind the president. President George H. W. Bush was about to announce something called the Space Exploration Initiative, or SEI. The very public nature of the ceremony was planned, and intended to kick-start the country on a new trajectory in space. America would explore again, and bring a permanent human presence to outer space. The forty-first president of the United States wanted SEI as part of his expansive goal to revive the nation's interest in math and science. He had a vision for the country for the last decade of the twentieth century: he wanted science, technology, engineering, and math education to reach every child in America.

One of Bush's first actions as president was to reestablish the National Space Council, as recommended by the Paine commission.[7] Stakeholders came from NASA, the CIA, Defense Department, Transportation, and OMB. Together, their mission was to develop for the president a coherent set of goals for the nation in space.[8]

For the civilian space program, the best thing about SEI was that it actually focused the nation's long-term goal in space: first settle the moon, then send an expedition to Mars. The thinking was that by revitalizing space exploration, America would gain momentum to increase its advantage in the world of high-tech. But SEI was actually much weaker than the Kennedy challenge thirty years earlier. The president called for the effort but did not commit the country to it. Political opponents, who viewed the Bush administration as nothing more than a third Reagan term, immediately criticized it as overly grandiose, vague, and lacking in clear direction.[9]

Bush responded by clarifying that there *was* a goal: Mars. He added a timetable to get there: "Thirty years ago, NASA was founded and the Space Race began, and thirty years from now, I believe man will stand on another planet. And so I am pleased . . . to announce a new age of exploration with not only a goal but also a timetable: I believe that before Apollo celebrates the fiftieth anniversary of its landing on the moon, the American flag should be planted on Mars."[10] The timetable thus became an integral part of the Bush initiative.

"If anyone doubted the president's strong commitment to a vigorous U.S. space exploration policy, this talk should straighten them out," said Paine. "Bush is displaying the vision and determination to get NASA moving outward again beyond Earth orbit."[11]

The previous spring, Paine and Quayle had talked about Mars. They discussed what could or could not be done within the confines of the SEI. To its credit, the White House was trying to, in some form, follow through on *Pioneering the Space Frontier.* He told Quayle that the NCOS recommendations were actually understood far better in 1989 than the goal of landing on the moon had been back in 1961. If the United States was to take the first steps to put human beings on Mars, it would serve as an immediate catalyst to revitalize the country's government-industry-university triad. Space would do its part to help sustain the economic recovery that had begun so well in the Reagan years. It would create opportunities for international cooperation. Education would be stimulated. Once a Martian community was established, in an early decade of the twenty-first century, future generations could continue the movement that would lead to discoveries to benefit the human race.[12]

But SEI gained no traction. The presidential initiative was the kind of action from the White House that NASA needed, but the onus was on the space agency to sell it and turn it into a workable program. Quayle wanted to know why NASA was getting bad press and why support was so tepid. Several internal NASA studies colored by the agency's own agenda had failed to turn up an answer. A frustrated White House then tasked Laurel Wilkening with providing an independent assessment. Her panel identified a foundational problem: President Bush had not adequately told the country *why* it should embark on a new space exploration program. Without a more compelling justification, the initiative could gain neither public support nor congressional funding.[13]

On a flight from the West Coast to Washington on July 11, 1990, Quayle asked Paine what he thought was happening. Paine couldn't wait to give the vice president his fully candid answer: The national media, for too long, had been overexuberant in their coverage of the space agency's woes. "Leaky shuttles, frisky astronauts, and blurry mirrors" had finally caught up with NASA, as he alluded to the media's overexuberant coverage of the agency's recent embarrassments. His position was that the space agency's public image would actually be a lot brighter if it would just admit its mistakes and promise to do better. An honest mea culpa was far better than a "stonewall, defense, siege" mentality. While the widely publicized problems with the Hubble space telescope and the personal indiscretions of a few astronauts did not require White House–level attention in 1990, two troubled programs did.[14]

One was the space shuttle. Paine had a strong position on the nation's flagship space program. First, continuing shuttle operations into the 1990s and beyond would hamper NASA's ability to carry out the Bush initiative. The shuttle was draining over half of NASA's entire budget every year. The agency simply could not branch out into something new as long as it had to keep the shuttles flying.

It also impeded the development of a new heavy-lift launch vehicle (HLLV). A new, next-generation rocket was needed to reduce the complexities, uncertainties, and, most importantly, cost of putting crew and cargo into low-Earth orbit. In fact, nothing in SEI could proceed until a solution was found to the problem of reliable, low-cost space transportation.

By 1990, the space shuttle had already been operating for nine years. It had been apparent after the first few missions that it could never provide low-cost access to Earth orbit. While it was an impressive, marvelous aerospace breakthrough and incredible flying machine, the likes of which the world had never seen before (or since), the shuttle also turned out to be the most expensive way to get into space.[15]

Congress had already approved $2 billion for a new orbiter (*Endeavour*) to replace *Challenger*. That was $2 billion which could have immediately gone into a new Mars program. While the shuttle proved to be a superb vehicle to periodically take six astronauts and a man-tended payload into orbit, it failed to reduce the cost of space transportation. The numbers did not lie: a manned shuttle flight to Earth orbit and back was more costly than a manned flight to the moon.[16]

Paine had the vice president's ear, and he told Quayle very candidly that the US needed to move beyond the shuttle as soon as possible. Space transportation using 1970s technology had simply become too risky and expensive. He asserted that NASA and the country could (and should) be rightfully proud of the space shuttle. But keeping the not-very-exciting "space truck with a mission no more glamorous than carting a load of toothpicks to Topeka" was confusing the means with the ends. The US should "declare victory and withdraw executive preoccupation with shuttle operations, substituting expendable launch vehicles wherever possible." Then, as the shuttle was being phased out, NASA and the Air Force should jointly develop a new-generation, man-rated rocket. He told Quayle that all shuttle-dependent programs needed to be reappraised. The sooner it was done, the higher the probability that a change would happen. That was the only way SEI could get off the ground.[17]

The other program that needed to go in a new direction was the space station. He had previously asked three presidents to approve a permanent space station in low-Earth orbit. In 1984, Reagan finally gave the go-ahead. But Paine and most experts familiar with the program were lukewarm at best to the space station that NASA called *Freedom,* because of its total dependence on the space shuttle. The agency had assumed that shuttle operations would be relatively inexpensive. But in order to construct and maintain a space station, it had to be very reliable and economical to operate (for fifty or more flights a year). Paine thus proposed that NASA redesign *Freedom* as an HLLV-launched spaceport in low-Earth orbit. The US could then place it in an orbit with the correct inclination (in the plane of the ecliptic) so it could serve as a gateway to launch human missions out into the inner solar system.[18]

His position was that NASA had totally missed the point of SEI, and along with it, a golden opportunity. The agency was too slow to respond to President Bush's initiative. It had been blinded by the day-to-day operation of the shuttle while trying to sell a space station to Congress. NASA had also failed to change with the times. The Soviet Union was now bankrupt, the president went out on a limb and proposed to the nation a very forward-looking space program with Mars as its goal, and yet NASA could not see itself beyond low-Earth orbit. Bush had announced a new initiative and then astonished even his critics by putting his reputation behind a substantial budget increase. While NASA welcomed the money, it gave the new goals low priority. And while policy-level communications between NASA and the White House had started out well, things regressed and eventually fell apart.

He told Quayle that there were just too many people with too many agendas. The best thing to do was to start over with an unbiased viewpoint. The White House needed a clear roadmap so its initiative could at least have a chance. Once that gained some traction, NASA could then take it and turn it into a history-defining program like Apollo. With nothing to lose, he boldly recommended to the White House that it could make a watershed decision by creating a national space exploration agency, or NSEA, from the ranks of the existing NASA infrastructure. It would then be up to the NSEA to carry out all the country's space exploration initiatives.[19]

Six days after his discussions aboard *Air Force Two*, Vice President Quayle told NASA Administrator Truly that he wanted a committee to objectively examine how NASA was carrying out the nation's long-term space policy. On July 25, 1990, the White House named Norman R. Augustine to lead a twelve-member Advisory Committee on the Future of the US Space Program. As the Chairman and CEO of Martin Marietta, Augustine was one of the most powerful executives in the aerospace arena and one of the most recognizable CEOs in the country. A few days earlier, he had called Paine to see if he would like to be on the committee. "Why not?" replied Paine from his office in Santa Monica.[20]

He wasted little time making his position known to his fellow committee members. A few days after the committee was formed, he articulated what he thought should be their priority: "I believe it is essential that we rapidly identify the half dozen key issues that we must address to help make the President's Space Exploration Initiative a historic triumph, arrange early inputs from America's leading experts in these fields, and concentrate our attention, talents and limited time on developing a few recommended high-level actions for our final report."[21]

But it wasn't his show to run this time. The Augustine committee differed from the NCOS five years earlier in several key respects. First, it had only 120 days to report back to the White House, instead of a year. Second, the underlying charter was different. While Paine's commission had been tasked to come up with a high-level vision for America in space, the Augustine committee had a much more specific focus. This time, the White House wanted an objective look at the nation's space policy. Its focus was on "problem solving" in order to save a stagnant space program. Both groups had to recommend a course for America in space, but the Augustine committee was formed specifically to address the difficulties of the SEI. It thus centered much more tightly and specifically on a way to get Bush's initiative started during his first term in office.[22]

Augustine immediately set a tone for the committee that was not what Paine had wanted or expected. In their first meeting, he turned his attention to earth and environmental sciences and deemphasized human spaceflight and space colonization. Augustine relegated those to an "open-ended schedule, tailored to match the availability of funds." Earth itself, and not outer space, became the focus. The committee, in effect, changed the spotlight of the US space program from "Mission from Planet Earth" to "Mission to Planet Earth."[23]

Paine did not agree with the change. He saw it as a clear deviation from the spirit of SEI. In his words, "the president's Space Exploration Initiative provides the greatest opportunity in 30 years to reinvigorate NASA through a forward-looking, purposeful, sustained civil space program."[24] Bush had given NASA a finite window of opportunity to initiate a Mars program. By shifting priorities, the committee was putting this golden opportunity at risk.

The disagreement was considerable. He and Wilkening (vice chair of the committee) were the primary opposition. Augustine was very keen on how the country's space program could be used as a way to help preserve the planet's environment. Global warming (climate change) was hotly debated, as were aspects of how the space program could be used to better manage natural resources here on Earth. "That was a pretty strong debate that we had," recalled Wilkening in hindsight. "Augustine's tack was that we need[ed] to worry about planet Earth."[25]

The country was not making much progress in outer space, and the committee was formed to put the program back on track. But even Paine had to admit that national enthusiasm for a long-term space program was at an all-time low. A month earlier, the Berlin Wall had been demolished. With the fall of the Soviet Union, the Cold War was now over. But Mars was still out there. Paine urged the committee to think of space as the next stepping-stone in the evolution of human civilization. His position never changed. Throughout that summer and into the fall, he spoke as much as he could to the national media, appeared on national television, and wrote articles making the case for Mars. The other members finally gave him a new nickname: "Chief Martian Monster."

He put the blame for NASA's inability to capitalize on SEI squarely on the shoulders of Dick Truly, the administrator of NASA. But as a member of the presidential committee guiding NASA, he feared that his outspokenness would end up doing more harm than good. He was also concerned that Truly might misconstrue his actions as being politically motivated. Nothing could be further from the truth. Privately, he told Augustine that "press reports of my critical concern over some NASA programs, buttressed by quotations from my letters, are embarrassing," and offered to "withdraw gracefully" from the committee.[26]

Augustine wouldn't hear of it, and rejected his talk of resigning. They came to an agreement. He would accommodate Paine's position, but would not adopt it as the central theme of the committee. They were, by now, also running out of time. The members went through three drafts of their report and met for a final session in Washington during Thanksgiving week 1990. Augustine told them that he wanted the report to be delivered in two weeks.

The timing was important. First, they had to meet Quayle's 120-day deadline. But as Augustine said, he also wanted to get the report into the hands of the White House and the right people in Congress before they left Washington for the Christmas holiday. Paine recalled spending that Thanksgiving writing the final report at home and in his Santa Monica office. In a conciliatory move, Augustine had asked him to write the committee's vision in an "inspirational statement of raison d'être." He wrote liberally and eloquently about why America was a stronger nation for having gone to the moon and why it could be stronger still by going to Mars. On December 9, the report was done.[27]

Quayle needed the committee to come through for him; Bush had given his vice president the responsibility to keep the exploration initiative alive. In a well-choreographed ceremony at the Old Executive Office Building on Monday, December 10, 1990, Augustine handed Quayle the commission's findings. Paine would call it only "concise and pragmatic."

Five themes stood out:

1. America should establish science as the highest priority of the civil space program. "Mission to Planet Earth" should be maintained at or above the current fraction of the space program's budget.

2. NASA should obtain exclusions for a portion of its employees from existing civil service rules. Or failing that, it should start converting selected field centers into Federally Funded Research and Development Centers affiliated with universities.

3. The US should redesign space station *Freedom* to reduce its complexity and cost. NASA should take whatever time was required to do this thoroughly and in a most innovative way.

4. NASA should pursue a "Mission from Planet Earth" as a complement to "Mission to Planet Earth," with Mars as the very long-term goal, but relieved of schedule pressures and progressing according to the availability of funding.

5. The US should reduce dependency on the space shuttle by phasing in a new, unmanned HLLV for all missions except those requiring a crew.[28]

The recommendations turned out to be not forward-looking at all, at least not in the way Paine wanted. But the White House could not wait any longer. It needed something tangible that Congress could approve quickly and NASA could then implement. Over his objection, the exploratory aspects that he championed were relegated to a low priority without a timeline. The ill-defined, long-term nature made it all that much more nebulous.

The overwhelming reaction from the space community, which had expected a set of forward-looking recommendations that could build on past achievements, was one of confounded disappointment. Said one national editorial, "Considering the makeup of the committee, there is no explanation but cowardice to account for how its members unanimously released a report which spits on the accomplishments of 30 years of man in space, and tells the young people of this country that there is no 'final frontier' for them to discover, explore, and develop." After Augustine and Wilkening presented the committee findings to Congress, a dejected Paine knew that they had missed out on a golden opportunity to move the country in space. "[I] hope your testimony today was better received than this," he told Wilkening[29]

Most demoralizing to him was that he knew he had lost on Mars.

EPILOGUE: A TWINKLE IN HIS EYE

In every battle there comes the time to fire your last torpedo.
—Tom Paine

On a restful evening in the spring of 1973, Tom Paine was on the telephone with the Reverend Francis B. Sayre Jr. of the National Cathedral. The fifth anniversary of Apollo 11's landing on the moon was still over a year away, but he was already planning ahead to something special. He was thinking of a unique way to mark the anniversary of that historic event.

When he and Barbara were in Washington, they had lived near the National Cathedral—a pillar for the common man, built to symbolize the profound faith in the Almighty upon which the country was founded. He had contacted Sayre with the idea to place a unique memorial in the cathedral. Perhaps, he proposed, some fitting work of art permanently enshrined in the hallowed halls by which visitors could perpend, reflect, and be inspired by.

Sayre, the first grandchild of President Woodrow Wilson, was the dean of the Washington National Cathedral for twenty-seven years. He embraced Paine's idea enthusiastically and personally took the proposal to the trustees of the cathedral. The trustees gave their unanimous approval for the project. Given the go-ahead, Paine commissioned a large stained-glass cathedral window for the center of the magnificent sanctuary's south wall. To get the project started, he donated $22,500 from his lecture honoraria.

In the next eight months, St. Louis artist Rodney M. Winfield designed and created a nineteen-foot glass work of art for the wall. From the time it was installed, the window has come alive each day under the bright rays of the afternoon sun. The brilliant colors give birth to an infinite realm of swirling stars and orbiting planets that reflect the human capacity for faith, the creation of the world, and the wonders of the cosmos. Winfield said that he had received his inspiration for the work from the photographs taken by the astronauts on their expeditions to the moon.[1]

"Once we were given the window, we began to wonder what it ought to show," said Sayre. "The idea occurred to me that whatever the window showed, it would be nice to have this fabulous artifact [a moon rock] embedded in the window."

Paine went to work. On November 2, he wrote President Nixon "to enlist your aid and participation" on behalf of the memorial in a unique way, by approving a piece of the moon to be prominently enshrined in the stained-glass window. Normally, NASA and the Smithsonian Institution decided on the disposition of the samples that were brought back from the moon. Paine bypassed that route and made a special appeal directly to the president.

With a piece of the moon permanently mounted inside the artwork, he hoped that a powerful and timeless symbol of human beings' first steps to the stars would be etched in a most compelling way. Its historical importance could not have been greater. He talked to James Fletcher and told Gene Emme, NASA chief historian, that "as a historian, you may understand why I am asking to divert a precious lunar sample from science to symbolism."[2] For assurance, Sayre pledged that the cathedral would take additional security measures to guard the sample.[3]

What worried the White House, though, was not security, politics, or religion. Rather, if it were to give a piece of the moon to the Washington Cathedral, it might have to do so for every church that asked for one. But the cathedral was not just any church. It was the national shrine as sanctified by its unique place in twentieth-century American history; nearly all national memorial services are still held there.

On January 14, Nixon gave his "delightful approval" for the work.[4] With the president's endorsement, the space agency transferred a two-inch-diameter piece of lunar sample brought back on Apollo 11 from the Lunar Receiving Laboratory at the Johnson Space Center to the cathedral. The 0.25-ounce piece of basalt from the Sea of Tranquility (Piece Number 230 from Rock 10057) was permanently embedded in a bubble at the center of the window.[5]

At 10:30 on the morning of July 21, 1974, the bells of the cathedral tolled loudly. Following Sayre to the pulpit, Paine read the Creation account from the book of Genesis. Fletcher followed with a reading of the account of Jacob's ladder to heaven. Neil Armstrong raised and presented the moon rock at the High Altar: "Very Reverend Sir, on behalf of the President and the people of the United States, we [the crew of Apollo 11] present unto you this fragment of Creation from beyond the Earth to be imbedded in the fabric of this house of prayer for all people." Canon Charles Martin, the headmaster of the Saint Albans School, pulled Paine aside and privately thanked him afterward for making the memorial "a wish come true."[6] The "Space Window" is now seen by over half a million visitors each year and is one of the most famous stained-glass works of art in the world.

Tom Paine had joined the space program at a time when America was impatient, restless, and waiting for something remarkable to happen. Each passing mission into outer space was making history. There is perhaps no better example of the value he placed on that history than the part he played in helping to bring about the Smithsonian's National Air and Space Museum.

In August 1969, he and Gene Emme revisited the long-standing plan to build a museum on the grounds of the National Mall. After World War II, the Congress and President Harry Truman had established a National Air Museum to house the country's many aviation artifacts that had been accumulating since before the turn of the century. Some dated as far back as 1876. The earliest were a group of exotic kites from the Chinese Imperial Commission given to America as a gift to celebrate the Centennial Exposition in Philadelphia. The museum did not have its own building then, so the eclectic collection of artifacts was temporarily housed in a room at the Smithsonian Arts and Industries Building.

Twelve years later, Congress authorized plans for a new National Air Museum with President Dwight Eisenhower's approval. With the dawn of the space age, President Johnson amended the legislation in 1966 to include the field of spaceflight. Architectural plans were drawn up and approved by the Smithsonian, but they lay dormant for the rest of the decade. With the costly conflict in Southeast Asia escalating, no appropriation was ever requested by Johnson before leaving office. The project became inactive, was put on hiatus, and soon "fell off the radar."

But now Nixon's enthusiasm was at a zenith following the moon landing. Paine capitalized on the opportunity and asked Emme to draft a letter to the White House. Would the president approve an "appropriate national living memorial commemorating the achievement of Apollo 11 [by] the construction of the National Air and Space Museum on the Mall?" He urged Nixon to seriously consider proceeding with this as a way to preserve the living legacy of flight. The nation, he said, deserved something beyond postage stamps and medals. In the same way, a stone structure, like the space monument in Moscow (the "Monument to the Conquerors of Space"), would be a rather cold, "sterile symbol." An inanimate object like that would just not be able to communicate the full and deep meaning of mankind's exploration of the stars. Landing on the moon was a product of American ingenuity, strength, and sacrifice made in a time of peace. The appropriate memorial should thus be a living institution. It should be a place where public support, education, and history could come together to celebrate mankind's glorious achievement in flight.[7]

Emme told him that they should, as much as possible, bypass the "clearly anti-historical" White House staffers and go directly to the president. Paine did not have to be reminded of this. Throughout the fall and into the following spring, they managed to avoid the gatekeepers of the Oval Office and worked directly to convince Nixon that the timing was right to request the $50 million that was needed. "Only with your personal support . . . has this any chance of mobilizing the support it will require." Paine suggested that the perfect date for the grand opening of the museum would be July 1976—the Bicentennial.[8]

This turned out to be the catalyst. Astronaut Mike Collins, who piloted command module *Columbia* in lunar orbit while Armstrong and Aldrin were on the surface, had recently retired from NASA. In 1971 the Smithsonian asked him to be the museum's first director. The following year, Congress appropriated $40 million for its construction. Things moved quickly from that point. Ground breaking took place on November 20 of that year. Artifacts for the "Milestones of Flight" gallery, including

Columbia, the *Wright Flyer*, Charles Lindbergh's *Spirit of St. Louis*, and Chuck Yeager's X-1 *Glamorous Glennis*, were carefully moved down the mall to the museum the following February. On July 1, 1976, President Gerald Ford presided over the ribbon cutting ceremony opening the doors of the National Air and Space Museum. Twoscore years later, the living memorial to the human conquest of flight is still the most visited museum in the world.

That summer, two very sophisticated robotic spacecraft called Viking 1 and Viking 2 landed on the surface of Mars after an eleven-month voyage in space. With the then-unheard price tag of $1 billion for two unmanned probes, the highly successful and highly publicized NASA missions seemed to rekindle the public's romance with the mysterious planet and the most tantalizing question of whether or not life existed beyond Earth. Futurists too looked for data that would support evidence of water and organic minerals, vital ingredients that would be needed for colonization.

Whenever there was talk of Mars, one did not have to look far to find him. It was about then that he began to network, in an informal way, with a group of space enthusiasts who called themselves the Mars Underground—an organization of Mars advocates started by students and scientists in Boulder, Colorado, who were quickly joined by peers from southern California. He shared a common vision with these scientists, students, and writers. They believed that Viking was not the end of Mars exploration but only the beginning of a journey that would be made first with robotic rovers, then by humans. He found the energy of the group's optimistic members refreshing and uplifting. Among them, he was not the administrator of NASA nor the president of Northrop, but just a fellow Mars advocate, one among many who came from all walks of life. It was just the way he wanted it.[9]

The Planetary Society, a more mainstream organization founded in 1980, allied with the enthusiasts to host a series of conferences on college campuses across the country. They wanted to make a case for the country to explore Mars. The Planetary Society believes that a good future is in store for human society. Their platform is that exploration that makes discoveries and creates new adventures beyond the current limits of technology is always a good thing. Their mission, to "empower the world's citizens to advance space science and exploration," points to Mars as the first stop.[10]

It didn't take long for him to make his mark. Paine wrote, researched, and lectured on how a settlement on Mars might work. He took things one step further than anyone had before and addressed the issues of governance and economy in a future Mars colony. The society welcomed his far-reaching, almost boyish enthusiasm.

Louis D. Friedman, a cofounder and former executive director of the society, recalls:

> We in the Planetary Society were almost neutral on the question of settlement. We were happy with the exploration of Mars and finding out what's there and finding out how suitable it was before we jumped into the next step of saying that we absolutely knew that settlement had to happen. We would argue about that sometimes. He would say we should go for Mars settlements and we would say no, we're for Mars exploration. But it was a give and take for people who really wanted the same thing, which was to get some vision going to Mars and learn more about it and gain that knowledge that would make us understand how to do it. . . . Our conversations were always about this notion of the goal of going further.[11]

Paine also had many conversations with physicist Gerard K. O'Neill. The late Princeton professor was a true futurist. He had postulated as far back as 1976 that humanity could achieve a good life, even at 20,000 times its current population, by expanding into space-based colonies. Like Paine, O'Neill wanted to follow the success of Apollo with grand endeavors in space to provide an option for the future of civilization. In his book *The High Frontier: Human Colonies in Space,* O'Neill had proposed that by moving society into space, human beings could increase cultural diversity and the quality of life. War and poverty would decrease here on Earth, and the population left on the planet might even dwindle to a pre-twentieth-century level. Large human colonies would populate the Earth-Moon system. The habitats would be made using materials mined from the moon and the near-Earth asteroids. Centrifugal force would create artificial gravity by spinning them at the right rate. Diurnal cycles replicating day and night, sunlight, and solar power would be generated from a host of accompanying solar satellites. Planet Earth would then become a great place to visit. But with more desirable options available elsewhere, many might choose not to live here.

Paine agreed with O'Neill's underlying vision. The previous August, he had written that a permanent settlement in space by the year 2075 was achievable. The current rate of technology growth could allow humanity to make such a leap into the solar system if necessary. He wrote O'Neill that technology would be shown one day to hold the key to the bondage of poverty, war, sickness, and overpopulation. Outer space held such promise that it could ultimately be what saves society from the planet in the event of a natural or man-made extinctive threat.[12]

Both men were simply weighing possibilities. Paine knew the difference between fact and fiction, between implementing and imagining. It was clear in his mind, however, that humanity in the last decades of the twentieth century was on the cusp of a new age of exploration not seen since Europe charted the Atlantic in the fifteenth century. Not all nations participated then, nor would they now, he said to an audience in London in 1972. "I believe we are in the initial stages of a new movement which will carry mankind beyond its present world view to a broader, three-dimensional perspective embracing all of the space within the Earth-Moon system and beyond."[13]

In his later years, a Boy Scout at a southern California troop meeting had asked him what he did in his life. He replied that he was a simple engineer who wanted to help make society a better place. He once contrasted himself to the mid-twentieth-century English essayist George Orwell. Both men speculated often on the future, but had come to very different conclusions. While both believed that technology was the engine for social change, the dark, foreboding fate of Orwellian humanity was the very antithesis of the bright prospect that Paine believed awaits mankind. Those who were trained in classics seemed to feel gloomy and overwhelmed, he once said, but those who are technically trained see challenge and believe they will meet it.[14]

After finishing his work on the Augustine committee in 1991, he kept up his busy pace. Arthur C. Clarke, Tom Clancy, and Ray Bradbury called him with their manuscripts and screenplays. Vice President Quayle asked him to critique the implications of the "Magna Carta of Outer Space" on the legality of the SDI missile defense program. The Marshall Space Flight Center asked him to help create the "History of Space Travel" display that welcomes visitors every day to the Huntsville, Alabama, space camp. The National Academy of Science's Naval Studies Board asked him to help locate the wreckage of the long-lost World War I Kaiserliche Marine (German Imperial Navy) SMS *Vulkan* U-boat salvage tug that sank in 1919. The US Navy asked him to help in the engineering research and development of the twin-hull ocean surveillance vessel design.[15]

After a brief illness, cancer took Tom Paine on May 4, 1992, at the age of seventy. Shortly before his death, he wrote that the days he spent in the Pacific had defined his life.[16] In 1944, before leaving on a war patrol, the future Barbara Paine had given him a book called *Stand By to Surface*. Written by the Australian author Richard Baxter, it told of the many daring exploits and missions of the Allied submarine forces during the war. The novel became the first of what would be a 3,400-volume collection of historical and scholarly materials on submarines and submarine warfare that he accumulated over the next forty-eight years. He had looked for books on the subject from all eras and especially those in languages other than English, and had found many written in Japanese, French, and Russian. Some dated to the late 1800s. After his passing, Barbara donated the tremendous collection to the Chester W. Nimitz Library at the United States Naval Academy. It was a way for his family to give something back to the country. Four years later, the donation was finalized; in 1997 the Navy held a special ceremony to dedicate the collection to all the men and women who have and will serve their country in the "silent service."[17]

On September 25 of that year, NASA launched the Mars Observer, the agency's first mission to the Red Planet since the Viking missions of 1975. On the side of rocket was a farewell inscription that read:

United States Spaceship *Thomas O. Paine*
Departed Earth September 1992.[18]

After sending the spacecraft off to Mars, the spent upper stage has been slowly, over time, captured by the pull of the sun. It now circles our sun on a timeless journey around the solar system.

Paine laid out a road map for human exploration of outer space, and that may still happen. When it does, pioneering colonists to the Red Planet will become the first Martians. They may even plant the Martian flag that Paine designed one night in 1984 when he perhaps had few better things to do than gaze up above the distant horizon. There he found an alluring speck of light awaiting him in the night sky over the dark expanse of the roaring Pacific. In the middle of the flag is a large red circle in the shape of the Roman symbol for Mars. It points upward toward a symbol that appears to be a star. In the opposite corner is Earth. While it is a simple piece of artwork, the message is unmistakable: Mars points the way to the galaxies.

"Tom Paine was a guy who just plain had his feet on the ground," recalls Jay Keyworth. "He grew up in a generation, as I did, where we all believed that technology could solve virtually any problem given enough time. He believed in big things being inspiring. . . . It was his nature to think big."[19]

Bud Bottoms, a Santa Barbara sculptor, was just a young artist starting out at TEMPO when he first met Paine. They became fast friends and talked often about the unlimited potential of humanity and the role that spaceflight would one day play. "He had a wild imagination," remembers Bottoms. Even back in 1963, "he needed an artist to make pictures of what he was thinking about, like life on Mars." When asked what he remembered most about his late, lifelong friend, Bottoms didn't hesitate. "He had a twinkle in his eye like you wouldn't believe. You always just knew something was going to happen when you were around him."[20]

AUTHOR'S NOTE

This was Arlington and here they all lay, formed up for the last time.
—James Salter

This biography of Tom Paine is a chronicle of his professional career, the narrative being how he left an indelible imprint on the US human spaceflight program. As is often the case, it is difficult to fully separate one's personal life from one's professional accomplishments. The story includes some, but not many, references to his personal life. The book is also not a management or space policy analysis. My aim was to tell the story of how one man's life played a memorable role in writing the ongoing history of the human conquest of space. His part in shaping those exciting events and his vision of what it truly means to the human race to be able to explore outer space are the focus of the story. I hope that readers find his journey fascinating, as I did, and that they will glean new information about the past that gives new perspective to today and tomorrow.

Tom Paine's career was quite well documented. Previously unpublished materials, archival collections, past and present oral history perspectives, and contemporaneous news articles were the primary sources for this biography. The preponderance of the material came from archival research. The Historical Reference Collection at NASA Headquarters in Washington, DC has an invaluable amount of material covering his years at the space agency and his involvement in the space advocacy movement. The Biographical, Administration and Organization, and Administrator Biography files in particular were foundational to the research. Likewise, the NASA Administrators Chronological Files from January 1968 through December 1970 were very valuable and served as a log of the day-to-day activities of his time at NASA. They provided a good accounting of what he did and did not do in his position as the head of the nation's civilian space program.

Another major archive was the collection of Thomas O. Paine Papers in the Manuscript Division of the United States Library of Congress. The amount of information—over 64,000 items in 183 containers—was imposing. I focused on the Personal Files, which contain some very useful information on his early years, school records, Navy records, and submarine service in the Pacific during World War II. Historical documentation of his career prior to and after NASA is well represented in the library's Corporate Files. They detail his work as an engineer at General Electric in Schenectady, Lynn, and Santa Barbara, and later his time as a high-level executive at GE and the Northrop Corporation.

Since his career took place before the digital age and the availability of electronic media, these files contain thousands of sheets of paper reports, hand-drawn plots, and notes written with pencils and fountain pens. Seeing the original handwriting brought another dimension of his work to life. It showed just how technical and mathematically complex Tom's work was as a young engineer. Mixed in with the files were personal notes that he wrote to himself. I found that these early files contrasted greatly with those from the later part of his career. The latter, which were mostly company-oriented records, were much more impersonal. Taken together, they formed a timeline of his career and preserved the details of a lifetime of work.

There were two collection of sources that I considered but did not use. One was the archives for General Electric TEMPO in Santa Barbara. TEMPO went out of business in 1980, and their files became (and remain) proprietary to GE. To that end, I cross-referenced what listings I found with what were openly available in the Library of Congress and decided that there existed enough material to paint a solid picture of that part of Tom's career.

The other was the full scope of his Navy work, some of which remains classified. Unlike NASA, which provides open and full access to researchers, the US Department of Defense does not operate in the same way. Here, I depended on Tom's own records, letters, notes, and other articles in open literature and decided this had to be sufficient for the scope of the book. As important as his naval service was, it was secondary to the main theme of this biography, that is, space exploration.

The Johnson and Nixon presidential libraries had some valuable information on Tom's interaction with the White House. These records document his meetings with Lyndon Johnson and Richard Nixon and provide a glimpse into what took place behind the scenes in those meetings.

Just as important were the oral history interviews that he gave. Robert Sherrod, the late war correspondent and journalist for *Time Life* magazine, and Gene Emme, NASA's first chief historian, conducted most of these around the time of the first moon landing, from 1968 through 1970. The interviews he did with Sherrod (who was an embedded journalist at NASA Headquarters during that time) were intended for a planned but never written book that would have chronicled the history of the Apollo lunar program. In addition to the interviews, Paine's many speeches, congressional testimonies, and writings provided insight given in his own words.

Other interviews conducted by historians with astronauts, managers, and his fellow luminaries in the fields of engineering and space exploration provided interesting and differing perspectives on what his contemporaries believed was most important about his legacy. Many of these are from the NASA Oral History Projects at the Johnson Space Center and at NASA Headquarters. They were invaluable sources that provided a spectrum of perspectives on Tom and his work. Taken together, these sources provided the body of evidence used to paint what I hope is an objective and compelling biography of the life and times of Tom Paine.

The book would not have come to fruition without the generous participation of many people. I am especially grateful to the distinguished colleagues of the late Dr. Paine who chose to share their recollections of him with me. Laurel Wilkening shared both serious and humorous stories of her time working with him on two presidential commissions and provided me with many useful notes and documents from those commissions. Thanks again, Laurel, for sharing those fond memories of Tom. I also thank Lou Friedman for conferring with me about his experiences and his many discussions with Tom about why we should never take our eyes off Mars. It was a personal privilege for me to interview Hans Mark, whose name I first heard as a young engineering student and someone whom I have long since come to appreciate as a giant of the space age. Thank you, Professor

Mark, for candidly conveying to me your perspective on the legacy of Tom. Jay Keyworth, I very much appreciated your perspective on how the Reagan White House viewed his recommendations for the future of the country in space and his vision for science and technology in America. Ed Hood helped me confirm some long-forgotten details of Tom's time at General Electric. And the late Neil Armstrong, who knew him from his days as an Apollo astronaut and later from their work together on the National Commission on Space, was very gracious to provide me with his perspective and help clarify details that only he could verify.

My utmost appreciation goes out to the fine professionals at Purdue University Press. I owe a large debt of gratitude to Peter Froehlich for taking a chance on the manuscript and recommending it to the press's editorial board for publication. I am very humbled that they selected it as the first title of the new Series on Aeronautics and Astronautics, and enthusiastically look forward to reading the great titles surely to come in the years ahead. Thanks also to Katherine Purple for overseeing the production of the book, turning it from paragraphs in a Word file into a complete publication. I very much appreciated the friendly approach and quick responses. And to Bryan Shaffer, thanks for getting the word out and promoting an interest in the work among your marketing partners. Thank you, Lindsey Organ, for creating such a vivid and handsome cover, and to Becki Corbin for making sure that I didn't forget anything along the way. A big "thank you" to the reviewers, copyeditor, and proofreader for all your knowledge and expertise. I truly appreciate it!

I also want to express my gratitude to my friends in the NASA History Program Office. Former Chief Archivist Jane Odom deserves a ton of credit for shepherding me through my NASA research. She was instrumental in pointing me in the right direction and helping me gain access to the right people to talk to at NASA and in the Manuscript Division of the Library of Congress. Chief Historian Bill Barry gave me candid and very helpful insight on what would make for a credible and objective historical biography of Administrator Paine. Much appreciation goes to Colin Fries. His tremendous knowledge of the archives at NASA Headquarters is unmatched and a great asset to anyone who wants to do research on NASA history and the space program. Thanks also to Gwen Pitman at NASA Media Services for her assistance in obtaining many of the photographs that are in the book.

Lastly, my appreciation to Johnny Tsiao, who took time out from his busy life to plod through and critique a very early version of the draft. What else are brothers for? Thanks, Johnny. As always, I owe my deepest gratitude to my family. Our dear Marissa, you especially encouraged me every time you asked, "Is that for the biography of Tom Paine?"

Thank you all.

SUNNY TSIAO

May 15, 2017

NOTES

The notes contain references to some correspondence and memoranda that may not be listed in the bibliography. Other archival materials were considered but not used, as discussed in the author's note.

PROLOGUE: MAN WILL CONQUER SPACE SOON

1. With Paine in the MOCR VIP viewing area were: Jim Elms, director of the NASA Electronic Research Center; Abe Silverstein, director of the Lewis Research Center; Rocco Petrone, director of launch operations at the Kennedy Space Center; Wernher von Braun, director of the Marshall Space Flight Center; his deputy, Eberhard Rees; Kurt Debus, director of the Kennedy Space Center; Edgar Cortright, director of the Langley Research Center; Charles Stark Draper, director of the MIT Instrumentation Laboratory; and Robert Seamans, secretary of the Air Force. There was also a large contingent of astronauts, including Apollo 10 commander Tom Stafford; Apollo 10 lunar module pilot Gene Cernan; Apollo 9 commander Jim McDivitt; and Mercury astronaut John Glenn. On the floor of the MOCR were a number of astronauts huddled around the capsule communicator (CAPCOM) console: Apollo 11 backup commander Jim Lovell and lunar module pilot Fred Haise; Apollo 12 commander Pete Conrad; Apollo 8 lunar module pilot Bill Anders; and rookie astronaut Charlie Duke, capsule communicator during the landing. Deke Slayton, director of flight crew operations at the Manned Spaceflight Center (MSC), was in the back row of the MOCR. With him were Bob Gilruth, director of MSC; Sam Phillips, director of the Apollo program; Chris Kraft, director of flight operations at MSC; and George Low, Apollo spacecraft program manager.

2. Neil A. Armstrong, e-mail message to author, March 8, 2011.

3. Thomas O. Paine, "Remarks at the Commonwealth Club" (speech, San Francisco, CA, August 1, 1969); Thomas O. Paine, interview by Robert L. Sherrod, September 10, 1970, transcript, 11.

4. Patricia N. Dolan, "Thomas Paine, Twentieth Century Version," Challenge, Fall 1969, 6.

5. Paine, "Notes for Speech on the Year 2000" (October 10, 1968).

6. Laurel L. Wilkening, interview by Carol Butler, November 15, 2001, transcript, 58.

1: NAVY BRAT

1. John R. Bartlett, "Records of the Colony of Rhode Island and Providence Plantations in New England," vol. 2, 564.

2. Paine, "Thomas O. Paine Current Biography," Personal Files, Thomas O. Paine Papers, US Library of Congress, Washington, DC (hereafter cited as LOC).

3. Ibid.

4. Ibid.

5. Thomas Buckley, "NASA's Tom Paine—Is this a Job for a Prudent Man?" *New York Times Magazine*, June 1969, 36.

6. Ibid., 37.

7. Personal Files, LOC; Carl Sagan et al., "The Last of the Corsairs: The Life of Thomas O. Paine," *Planetary Report* 12, no. 5, (September/October 1992).

8. Personal Files, LOC.

9. Paine was convinced that he ended up studying harder and reading more at Brown than he would have at Annapolis, and that when he graduated, his eyes were in no worse shape than when he entered college.

10. Paine, interview by Sherrod, October 7, 1971, transcript, 14.

11. Personal Files, LOC.

2: I NEVER GOT OVER IT

1. Naval service records, Personal Files, LOC.

2. Ibid.

3. Naval service records; Thomas Buckley, "NASA's Tom Paine: Is This a Job for a Prudent Man?" *New York Times Magazine*, June 1969, 36; Paine, "Sharply Etched Unforgettable Images from the World War II Pacific Submarine War," Personal Files, LOC.

4. Paine, "Sharply Etched."

5. Naval service records.

6. Paine, "Sharply Etched." Besides Thomas Paine, names of crewmembers with historical counterparts on the USS *Pompon* included a Benjamin Franklin III and a cook named George Washington.

7. Paine, "Southwest Pacific 1943," wartime journal, Personal Files, LOC, 24–25; Paine, "Sharply Etched."

8. Ibid, "Southwest Pacific."

9. Paine, "Southwest Pacific," 6, 16–17, 27.

10. Lieutenant Commander Stephen H. Gimber had relieved Commander Hawk at the end of the third war patrol.

11. Bernie Kanoff, "USS Pompon (SS-267): A History of Personnel and War Patrols," http://www.homepages.donobi.net/bkanoff/pompon/pompon.html; Paine, "U.S.S. Pompon (SS267) Fifth War Patrol off the Japanese Coast" and "U.S.S. Pompon (SS267) Report of the Fifth War Patrol," Personal Files, LOC.

12. Ibid.

13. A. T. Lawton, "Dr. Thomas Paine—An Appreciation," *Spaceflight* 34, September 1992.

14. Paine, "U.S.S. Pompon (SS267)-Report of the Sixth War Patrol," Personal Files, LOC; Kanoff.

15. Buckley, "NASA's Tom Paine," 36.

16. Kanoff; Paine, "Sharply Etched"; Naval service records; "Citation to Lieutenant (JG) Thomas O. Paine, United States Naval Reserve from C. W. Nimitz," Personal Files, LOC.

17. Kanoff; Paine, "Sharply Etched."

18. Paine, "Southwest Pacific."

19. Paine, "Wartime Journal," July–December 1945, 10.

20. Commander, Submarine Force, US Pacific Fleet Public Affairs, Story Number NNS060616.16, June 16, 2006, http://www.navy.mil/submit/display.asp?story_id=24211.

21. Paine, "Report of the Fifth War Patrol."

22. Paine, "Wartime journal," 17.

23. Paine, "Sharply Etched."

3: A LONG VOYAGE HOME

1. The *Peaceful P* earned four Battle Stars for valor during World War II. She was decommissioned the following year and recommissioned in 1953 as the radar picket vessel SSR-267. The boat went on to serve another six years supporting NATO and the Sixth Fleet in the Mediterranean, Atlantic, and Caribbean. A propeller from the *Pompon* is on loan from the Naval Historical Center and on display today overlooking the west bank of the Potomac River in a city park in Old Town Alexandria, Virginia. (George Paine, "USS Pompon [SS-267]," http://www.pacerfarm.org/i-400/ss267.htm.)

2. Paine, "Wartime journal"; Paine, "I Was a Yank on a Japanese Sub," *IEEE Proceedings,* September 1986, 73–78.

3. Paine, "Sharply Etched."

4. The I-400 was an incredible submarine. She had on board 213 crewmen when she was intercepted by allied forces on August 28, 1945, just east of Honshu, Japan. A Japanese officer told Paine that it had carried as many as 220 men to facilitate rapid submarine and aircraft operations at sea. It could launch three planes hangared in its hull in succession within 45 minutes of surfacing.

5. Commanding another captured super-sub, the I-14, was Commander John S. (Jack) McCain Jr. Paine would meet him again twenty-two years later while working on the Apollo program. McCain was the commander in chief of the US Pacific Fleet (CINCPAC), responsible for recovering the astronauts after they splashed down in the Pacific. His son, Senator John S. McCain III, would win the nomination of the Republican Party for president in 2008.

6. Telegraph dated December 11, 1945, Personal Files, LOC.

7. The Hawaii Undersea Research Laboratory at the University of Hawaii found the I-400 in 2,300 feet of water off the southwest coast of Oahu in August of 2013. The wreckage of her scuttled sister sub, the I-401, had also been rediscovered in March of 2005 by the laboratory during a training exercise. It was still sitting upright at a depth of 2,850 feet off Barbers Point, Oahu.

8. Paine, Letter to the Chief of Naval Personnel via the Commanding Officer USS *Pompon*, Personal Files, LOC.

9. Worksheet of J. M. McDowell, Personal Files, LOC.

10. Paine, handwritten letter to home, Personal Files, LOC.

11. Paine, Naval service records, Personal Files, LOC.

12. Miscellaneous notes, Personal Files, LOC.

13. Paine, letter to home, July 29, 1945, Personal Files, LOC.

14. Paine, interview by Arthur Garratt, June 9, 1975, transcript, 2; Anne Woodham, "To Outer Space from Australian Island," *Australian Women's Weekly*, July 16, 1969, 5.

4: HOUSE OF MAGIC

1. Stanford University papers, Personal Files, LOC.

2. Ibid.

3. He held a Secret clearance with the Department of Defense and a Q clearance (equivalent to Top Secret) with the Department of Energy.

4. Stanford University papers.

5. Paine, letter to O. C. Shepard, August 6, 1950, and resume from September 1962, Personal Files, LOC.

6. Ibid.

7. Paine, resume from 1962.

8. Paine, Shepard letter.

9. Space World, October 1987, 12.

10. Paine, miscellaneous papers, Corporate Files, LOC.

11. W. Henry Lambright, Powering Apollo: James E. Webb of NASA (Baltimore: Johns Hopkins University Press, 1998), 192.

12. TEMPO also had small offices in Honolulu and in Washington. The name originally stood for Technical Military Planning Operation. Its first employees nicknamed it GE's Temporary Office. The name stuck and quickly became the official business name with no specific meaning.

13. "1980," Talk of the Town, *New Yorker*, May 22, 1965, 37–38.

14. Buckley, "NASA's Tom Paine," 37.

15. Paine, resume, September 1962.

16. Tyler Blue, "Houston, We've Had a Problem," Santa Barbara News-Press, April 18, 2010, http://tylerblue.com/2010/07/Houston-weve-had-a-problem.

17. Buckley, "NASA's Tom Paine," 38.

18. "1980."

19. Paine, interview by Sherrod, September 10, 1970, transcript, 42–45.

5: IT'S A PRESIDENTIAL APPOINTMENT?

1. The Apollo spacecraft had three main modules. The command module (CM) was the cone-shaped capsule that housed the crew of three astronauts. With an interior volume of just over two hundred cubic feet, it was about as roomy as a modern-day minivan. But in the weightlessness of space, the crew could utilize the entire three-dimensional volume. The CM was the only part of the entire launch vehicle and spacecraft assembly that was recovered at the end of a flight. Right behind it was the service module (SM) that contained the power generation subsystems, oxygen tanks, fuel tanks, and the large service propulsion system

rocket engine. The SM was jettisoned just before the spacecraft reentered Earth atmosphere. Together, the two modules were referred to as the CSM. The third piece of hardware was the lunar module (LM). Upon arrival in lunar orbit, the LM separated from the CSM to land two astronauts on the surface. The LM had two stages. The descent stage powered the craft down to the surface while the ascent stage returned the crew to lunar orbit. It was jettisoned before leaving lunar orbit.

2. "Loose Leadership," *Aviation Week and Space Technology*, April 17, 1967, 25.

3. William J. Normyle, "NASA Shifts Key Apollo Officials," *Aviation Week and Space Technology*, April 10, 1967, 26.

4. James M. Beggs, interview by Kevin Rusnak, March 7, 2002, transcript, 82.

5. Robert C. Seamans Jr., *Aiming at Targets*. (Washington, DC: NASA, 1996), ch. 4.

6. Ibid.

7. Macy had also served as the executive director of the Civil Service Commission under President Eisenhower.

8. Buckley, "NASA's Tom Paine," 41.

9. Ibid., 43.

10. Unlabeled note paper, Folder 49901, NASA Historical Reference Collection, NASA History Program Office, NASA Headquarters, Washington, DC.

11. Buckley, "NASA's Tom Paine," 41.

12. Lambright, *Powering Apollo*, 191–192.

13. Ibid.

14. Lyndon B. Johnson, "Daily Diary January 31, 1968," Daily Diary Collection, LBJ Presidential Library, Austin, Texas, 3.

15. Thomas O. Paine, interview by T. Harri Baker, March 25, 1969, transcript, 1–2; Paine, interview by Sherrod, August 14, 1970, 22.

16. Glen E. Swanson, *Before This Decade Is Out . . . Personal Reflections on the Apollo Program* (Washington, DC: NASA, 2004), 28.

17. Thomas O. Paine, interview by Eugene M. Emme, July 9, 1970, transcript, 2.

18. Paine, interview by Sherrod, August 14, 1970, 15.

19. Paine, interview by Sherrod, September 10, 1970, 7; Lambright, *Powering Apollo*, 5–6.

20. Wilkening, interview by Butler, 22.

21. Paine, interview by Sherrod, September 10, 1970, 7–9.

22. Thomas O. Paine, "Summary of Remarks to the Management Advisory Panel," speech, Washington, DC, May 14, 1968.

23. Frank J. Magliato, "Note to Mr. Dembling," correspondence at NASA Headquarters copied to Paine, May 8, 1968.

24. Buckley, "NASA's Tom Paine," 43.

25. Ibid.; Paine, interview by Emme, 6.

26. Ibid.

27. "Press Briefing with George Christian and James Webb," transcript of White House press conference, Washington, DC, September 16, 1968.

28. Paine, interview by Baker, 6–7.

29. Only later was Paine formally referred to on record as the NASA acting administrator. This detail regarding his official title only came to light some two and a half months later when the NASA Office of Public Affairs and the White House wanted clarification under the intense media scrutiny surrounding Apollo 8. (Unlabeled handwritten note on Paine's formal title, December 20, 1968.)

30. Paine, interview by Baker, 9.

31. Ibid., 11–13.

6: GAINING SOME RESPECT

1. The Apollo-Saturn flights were renumbered after the January 1967 launch pad fire. AS-204 was renamed Apollo 1 in honor of Grissom, White, and Chaffee. There was no Apollo 2 or Apollo 3. Apollo 4, on November 9, 1967, was the first unmanned flight test of the Saturn V launch vehicle. This was followed by an unmanned test of the lunar module in low-Earth orbit on Apollo 5, January 22, 1968. Apollo 6 was another unmanned flight test of the Saturn V on April 4, 1968. This was then followed by the first manned flight Apollo 7 in October.

2. NASA had started on CSM Block II improvements before the AS-204 accident, but the pace at which redesigns took place increased dramatically after the fire.

3. Thomas O. Paine, "Speech to Aerospace Industries Association," presentation, Phoenix, AZ, November 20, 1968.

4. Paine, interview by Baker, 44.

5. Courtney G. Brooks, James M. Grimwood, and Loyd S. Swenson, Jr., *Chariots for Apollo: A History of Manned Lunar Spacecraft* (Washington, DC: NASA, 1979), 234–235.

6. Paine, interview by Sherrod, January 23, 1969, 38.

7. Brooks et al, *Chariots for Apollo*, 258–259; "Minutes of Meeting to Review Technical Feasibility of AS-503/CSM Only Mission" (NASA Headquarters, Washington, DC, August 14, 1968). Apollo 7 had not yet launched at the time of the meeting.

8. Edgar M. Cortright, *Apollo Expeditions to the Moon: The NASA History* (Mineola, NY: Dover Publications, 2009), 172.

9. Buckley, "NASA's Tom Paine," 44.

10. Ibid.

11. Cortright, *Apollo Expeditions to the Moon*, 172–175.

12. Paine, interview by Sherrod, January 23, 1969, 28.

13. Jim McDivitt, Dave Scott, and Rusty Schweickart were on the schedule to fly before Frank Borman, Jim Lovell, and Bill Anders. The Apollo 8 decision swapped the flight order of the two crews.

14. Paine, interview by Sherrod, May 13, 1969, 31.

15. George E. Mueller, "Response to Question on Apollo Service Propulsion System Engine," letter to NASA Acting Administrator, NASA Headquarters, Washington, DC, November 20, 1968).

16. Thomas O. Paine, "Letter to Dr. Eberhard F. M. Rees on his retirement from NASA," personal correspondence, January 22, 1973.

17. Brooks et al, *Chariots for Apollo*, 258–259; Administrator Chronological Files, August 1968 (NASA History Program Office, NASA Headquarters, Washington, DC).

18. Paine, interview by Sherrod, January 23, 1969, 28.

19. Paine, interview by Baker, 62.

20. Robert Zimmerman, *Genesis: The Story of Apollo 8* (New York: Random House, 1998), 197.

21. Borman turned Paine down, however. Jim Lovell, who wanted a chance to walk on the moon, also turned down the job. Apollo 9 Commander Jim McDivitt had expressed some interest but wanted to finish his career in the Air Force and continue working on the Apollo program. In the end, Bill Anders accepted Paine's recommendation and Agnew's nomination to head the Space Council. (Paine, interview by Sherrod, May 13, 1969, 48).

22. Thomas O. Paine, Transcript of proceedings from an awards presentation at the Franklin Institute, Philadelphia, PA, March 4, 1969.

23. Thomas O. Paine, "Letter to William C. Schneider on his retirement," personal correspondence, March 14, 1980; Thomas O. Paine, "Letter to the Editor: USSR Views Apollo 8," *Aviation Week and Space Technology*, January 13, 1969, 12.

24. In the post-flight awards ceremony honoring the astronauts, Paine, who was just behind Dean Rusk as they walked into the East Room of the White House, heard the Secretary of State murmur, "My God, I never thought we'd see this day."

25. Paine, interview by Baker, 59.

26. Christopher C. Kraft Jr., interview by Rebecca Wright, May 23, 2008, transcript, 20.

7: I'M NOT A POLITICIAN

1. Paine said after the election that had Hubert Humphrey or Robert Kennedy won, it was his understanding that retired astronaut John Glenn would have been the leading candidate to be NASA administrator.

2. "President Must Soon Appoint Space agency Administrator," *Washington Post*, February 9, 1969, A11.

3. Robert Sherrod, handwritten note in Paine's Administrator files, December 17, 1971.

4. William Hines, "Retired Air Force General May Head NASA," *Washington Star*, November 17, 1968, F4.

5. *Washington Post*, February 9, 1969, A6.

6. Paine, interview by Sherrod, January 23, 1969, 1.

7. Seamans, *Aiming at Targets*, 58.

8. Harry S. Flemming, "Memorandum for the President: Nomination of Thomas O. Paine of California as Administrator, National Aeronautics and Space Administration," White House, Washington, DC, March 5, 1969.

9. Lovell accepted the award on behalf of Borman and Anders, who were both in Houston supporting Apollo 9.

10. "The Best Man," *Newsweek*, March 17, 1969, 74.

11. "Exchange of Remarks between The President; James M. Murray, President, National Space Club; Mrs. Robert H. Goddard; Captain James A. Lovell, Jr., and Thomas O. Paine, Acting Administrator of NASA, at the Ceremony Awarding the Robert H. Goddard Memorial Trophy to the Apollo 8 Astronauts," Office of the White House Press Secretary, March 5, 1969.

12. NASA Headquarters Weekly Bulletin, Washington, DC, April 3, 1969.

13. Paine, interview by Sherrod, August 14, 1970, 31.

14. Ibid., 31–32.

15. Josie A. Soper, interview by Rebecca Wright, April 19, 2006, 18–19.

16. Thomas O. Paine, "Statement before the Committee on Aeronautical and Space Sciences," US Senate, Washington, DC, October 3, 1968; NASA Administrator Chronological Files, March 1969, NASA History Program Office, NASA Headquarters, Washington, DC.

17. Thomas O. Paine, draft preface to the supplement of the preliminary history of NASA during the administration of Lyndon B. Johnson.

18. Buckley, "NASA's Tom Paine," 34.

19. "Paine Says NASA Can Justify $5–$6 Billion Budget," *Space Daily,* June 9, 1969, 172.

20. Paine, "Statement before the Senate Committee on Aeronautical and Space Sciences," US Senate, April 6, 1970; Administrator Chronological Files, February 1969; Thomas O. Paine, "Remarks at the Apollo History Workshop," speech, Washington, DC, May 21, 1970.

21. Thomas O. Paine, "Space Age Management and City Administration," *Public Administration Review,* November/December 1969, 7–14.

22. Paine, "Speech to Aerospace Industries Association."

23. Paine, "Summary of Remarks to the Management Advisory Panel."

24. "Interview with Thomas O. Paine, Head of National Aeronautics and Space Administration," *U.S. News and World Report,* July 7, 1969, 28–33.

25. David DeVoss, "So Says Thomas O. Paine," *Los Angeles Times Magazine,* April 13, 1986, 23–25; Paine, "Speech to Aerospace Industries Association."

26. Apollo 9 provided history with a very iconic NASA photograph. The first onboard picture taken by Lunar Module Pilot Rusty Schweickart from outside the lunar module showed Command Module Pilot Dave Scott wearing a bright red helmet and standing up through the hatch opening of the command module. Taken from high above the atmosphere, the detail, clarity, and sweeping panorama of the picture were most impressive. It delighted Paine; upon seeing it after the mission, he ordered copies sent to dignitaries and officials around the world. (Paine, miscellaneous notes, Government Files, LOC.)

27. Robert C. Seamans Jr., letter to Paine, March 18, 1969

28. Paine, interview by Emme, September 3, 1970, 22–23.

29. Brooks, Grimwood, and Swenson, *Chariots for Apollo*, 300–302.

30. The Lunar Landing Training Vehicle (LLTV), nicknamed "The Flying Bedstead" by pilots and technicians, was a unique, open framework, flying vehicle used by astronauts to simulate the final phase of a soft landing on the moon. A gimbaled turbofan jet engine could be throttled to support five-sixths of the vehicle's weight to simulate the reduced lunar gravity, while two thrusters simulated the LM descent stage engine. The LLTV proved to be a valuable training tool but was very complicated and highly unstable. Three vehicles were lost in accidents (all three pilots ejected safely, including Neil Armstrong) due to malfunction of the complex onboard thrust control system.

31. Paine, interview by Sherrod, May 13 1969, 41–42.

8: A GREAT SENSE OF TRIUMPH

1. "NASA Chief Hits Talk by Kennedy," *Washington Post,* May 21, 1969, A2.

2. Paine, interview by Emme, 3 September 1970, pp. 4–7.

3. Ibid., 7-10. A 160 gram (5.6 ounce) piece of lunar breccia brought back to Earth by the crew of Apollo 15 in August of 1971 is on display today at the Kennedy Library and Museum in Boston, Massachusetts. After years of construction setbacks, the library was finally dedicated and opened to the public in October 1979.

4. The lunar module stowed inside the Saturn V launch vehicle could be accessed for last-minute items and changes while it was sitting on the launch pad through a panel hatch in the side of the spacecraft LM adaptor (SLA).

5. A sidebar on the flag that rests on the Sea of Tranquility: it is a $5.50, off-the-shelf, store-bought flag measuring approximately 2.5' x 4'. It was purchased at a Sears department store at the Baybrook Mall, just down NASA Road 1 from the Manned Spacecraft Center. A government GSA purchase order had been written for the flags, but time had run out. Technicians had to start working on one so it could be shipped down to the Kennedy Space Center and stowed on the lunar module in time for the launch. After secretaries, over their lunch hours, brought three flags back from the mall, workers in the Technical Services Division of MSC embedded one with a telescoping rod across the top so it could be deployed on the moon. (Author's recollection of conversation with JSC engineer

Carl Hohmann; Anne M. Platoff, "Where No Flag Has Gone Before: Political and Technical Aspects of Placing a Flag on the Moon," NASA Contractor Report 188251, October 11, 1992.)

6. Hans M. Mark, interview by author, January 29, 2011, transcript, 7.

7. Paine, interview by Sherrod, August 14, 1970, 34–35.

8. Ibid. The other piece of historical accoutrement left on the surface was a small silicon disc with goodwill messages from US Presidents Eisenhower, Kennedy, Johnson, and Nixon; leaders of seventy-three other countries; and Pope Paul VI. The State Department had helped NASA obtain the goodwill recordings from the world leaders. The names of high-level NASA managers, along with the names of congressional leaders, were also inscribed. The fragile disc was carried to the moon inside a protective aluminum capsule and carefully placed on the surface by Armstrong during the extravehicular activity.

9. Paine, "Remarks at the Commonwealth Club."

10. Paine, interview by Emme, September 3, 1970, 2–4.

11. Ibid., 6.

12. Thomas O. Paine, "Address before the National Press Club," speech, Washington, DC, August 6, 1969.

13. Apollo 10's Stafford, Young, and Cernan had all flown previously on Project Gemini. Tom Stafford had already logged over ninety-eight hours in space and had commanded Gemini 9. John Young had flown on the first Gemini flight in 1965 and later commanded Gemini 10. Gene Cernan had been Stafford's crewmate on Gemini 9. Neil Armstrong had previously commanded Gemini 8, while Mike Collins was the pilot on Gemini 10. Buzz Aldrin flew on Gemini 12, the final flight of the program, and logged over ninety-four hours in space. While Alan Bean would make his first flight into space on Apollo 12 in November 1969, Pete Conrad had flown on Gemini 5 and then commanded Gemini 11, and Command Module Pilot Richard Gordon had been the pilot on Gemini 11. Stafford would later go on to command the Apollo-Soyuz Test Project in 1975, while Young and Cernan would walk on the moon in 1972 as commanders of Apollo 16 and 17, respectively. Following Apollo 12, Conrad and Bean commanded the first two manned Skylab missions in 1973.

14. Armstrong, e-mail to author, April 2, 2011; Brooks, Grimwood, and Swenson, *Chariots for Apollo*, 320–322.

15. Armstrong, e-mail to author, April 2, 2011; Cortright, *Apollo: Expeditions to the Moon,* 146–150, 160.

16. NASA Administrator chronological files, January–September 1969; James R. Hansen, *First Man: The Life of Neil A. Armstrong* (New York: Simon & Schuster, 2005), 368–373.

17. The third member of the crew, Command Module Pilot Mike Collins, was in orbit around the moon.

18. In 2012, after Neil Armstrong's passing, his brother Dean revealed in an interview that Armstrong had prepared his historic quote and shared it with him several weeks prior to leaving for the moon. (Richard Gray, "Neil Armstrong's family reveal origins of 'one small step' line," *The Telegraph*, December 30, 2012, http://www.telegraph.co.uk/news/science/space/9770712/Neil-Armstrongs-family-reveal-origins-of-one-small-step-line.html).

19. Thomas O. Paine, "Anniversary Story," *Die Welt,* July 1969, 5.

20. "Interview with Thomas O. Paine" *U.S. News and World Report*, July 7, 1969, 28–31.

21. Thomas O. Paine, "Address to Australian Ministers and Officials," speech, Canberra, Australia, February 26, 1970.

22. Paine, "Remarks at the Commonwealth Club"; Paine, "Address before the National Press Club."

23. Paine, "Anniversary Story."

24. Paine, interview by Sherrod, September 10, 1970, 15–16.

25. Ibid.

26. Paine, interview by Emme, August 3, 1970, 27–28. On May 1, 1960, a CIA U-2 reconnaissance aircraft was shot down over the Soviet Union, two weeks before the scheduled opening of an East-West summit in Paris with the United States, the Soviet Union, the United Kingdom, and France. The international incident collapsed the talks and signified a major deterioration in US-USSR relations at the end of Eisenhower's presidency.

27. Paine, "Anniversary Story."

28. Paine, "Address before the National Press Club."

29. At the time Apollo 8 orbited the moon in December of 1968, Paine gave them eighteen months to land a cosmonaut on the surface; Webb thought that it would happen much sooner. The Soviets never made it to the moon.

30. Paine, interview by Emme, September 3, 1970, 23–25.

31. Paine, "Summary of Remarks to the Management Advisory Panel."

32. Paine, interview by Emme, July 9, 1970, 33; Paine, "Statement before the Senate Committee on Aeronautical and Space Sciences," April 6, 1970.

33. New York City Public Events Commissioner John S. Palmer estimated the parade crowd at 4 million. Other estimates were lower, and were based on the fact that the VIP party was more than one hour ahead of schedule. (Joseph Lelyveld, *New York Times*, August 14, 1969, 1.)

34. Don Oberdorfer, *Washington Post,* August 14, 1969, A1.

35. Julie Nixon Eisenhower, *Pat Nixon: The Untold Story* (New York: Simon & Schuster, 1986), 268–272.

36. Thomas O. Paine, "Report on the Flight of Apollo 11 by Dr. Thomas O. Paine, Administrator of the National Aeronautics and Space Administration, to the United Nations Committee on the Peaceful Uses of Outer Space," NASA Press Release USUN-94(69), Washington, DC, September 8, 1969.

9: GOING GLOBAL

1. Paine, "Address to Australian Ministers and Officials."

2. Arnold W. Frutkin, interview by Rebecca Wright, March 8, 2002, transcript, 1–3.

3. Paine, "Address to Australian Ministers and Officials."

4. Thomas O. Paine, "Speech before the Air Force Association," speech, Houston, TX, March 20, 1969.

5. Paine, "Statement before the Senate Committee on Aeronautical and Space Sciences," March 11, 1970.

6. Frutkin, interview by Wright, 1–3.

7. ESRO and ELDO merged in 1975 to form what is now the European Space Agency, or ESA.

8. Paine, "Statement before the Senate Committee on Aeronautical and Space Sciences," March 11, 1970.

9. Paine, interview by Emme, July 9, 1970, 5.

10. In December 1966, NASA had launched the first of three successful Application Technology Satellites (ATS) that experimented with new ways to relay information from geosynchronous orbit down to Earth. Since a satellite in

geosynchronous orbit stays nearly over the same ground location on the Equator as it circles the globe, it is perfect for relaying communications and television signals. India was a good candidate for the proof-of-concept data relay experiment, as NASA could maneuver the ATS and place it over the large landmass of India. SITE, or Satellite Instructional Television Experiment, used the ATS to transmit television signals to the entire Indian subcontinent for the first time ever.

11. Frutkin, interview by Wright, 22–25; NASA News Release 69-135, NASA Headquarters, Washington, DC, September 18, 1969; "Memorandum of Understanding Between the Government of India and the Government of the United States of America Regarding India-U.S.A. ITV Satellite Experiment Project," http://www.Indianembassy.org/indusrel/India_US_treaties/satellite_september_18_1969.html.

12. Hans M. Mark, interview by the author, January 29, 2011, transcript, 9.

13. Edward C. Ezell and Linda Neuman Ezell, *The Partnership: A History of the Apollo-Soyuz Test Project* (Washington, DC: NASA, 1978), 1.

14. Thomas O. Paine, "Letter to Mr. Edward C. Ezell," personal correspondence, July 12, 1974.

15. Paine, "Statement before the Senate Committee on Aeronautical and Space Sciences," April 6, 1970.

16. Frutkin, interview by Wright, March 8, 2002, 45.

17. Paine, "Remarks at the Apollo History Workshop."

18. DeVoss, "So Says," 23–25.

19. Ezell and Ezell, *Partnership,* 11.

20. Mathematician Keldysh (1911–1978) was the behind-the-scenes counterpart to the imposing Sergei Korolev, remembered now as the chief designer of the Soviet space program.

21. Ezell and Ezell, *Partnership*, 3–4.

22. Thomas O. Paine, "US/USSR Cooperation in Space Research," Attachment E of Statement before the Senate Committee on Aeronautical and Space Sciences, March 11, 1970.

23. Frutkin, interview by Wright, March 8, 2002, 47.

24. Roland W. Newkirk, Ivan D. Ertel, Courtney G. Brooks, *Skylab: A Chronology* (Washington, DC: NASA, 1977), 184.

25. Paine, "Remarks at the Apollo History Workshop."

26. Ezell and Ezell, *Partnership,* 11. Other ideas were more tangential. These included a unified network of tracking stations, exchange of scientific data from space experiments, and joint use of meteorological satellites.

27. Edward H. Kolcum, "U.S., USSR Set Joint Space Talk," *Aviation Week and Space Technology*, October 19, 1970, 16–17.

28. Robert Gilruth, Glynn S. Lunney, and Caldwell C. Johnson led the team from the Manned Spacecraft Center.

29. With advances in telecommunications technology in the six years after 1969, ASTP was broadcast to a live television audience much larger than even the Apollo 11 lunar landing mission. It was the last Apollo spacecraft to fly into space and would be the last expendable, manned US space capsule to splashdown in the Pacific in the twentieth century. The three American astronauts were in orbit for nine days while the two Soyuz cosmonauts stayed for six.

30. James C. Fletcher, "Letter to the Honorable Thomas O. Paine," June 27, 1972.

10: WHAT NOW?

1. Paine, Statement before the Senate Subcommittee on Independent Offices, Committee on Appropriations, May 19, 1970.

2. Zack Strickland, "NASA Reduces Manned Space Program," *Aviation Week and Space Technology,* September 7, 1970, 18–19.

3. Arthur J. Snider, "NASA Not Seeking to Be 'Czar of Science'-Paine," *Chicago Daily News,* December 27–28, 1969, 9.

4. Ibid. Cosmonaut Georgi T. Beregovoi told Paine that Russian scientists and engineers were ferocious toward each other and, unfortunately for their space program, the scientists had been winning.

5. Dolan, "Twentieth Century Version," 4.

6. Paine, interview by Sherrod, October 7, 1971, 20.

7. Paine, Statement before the Committee on Aeronautical and Space Sciences, June 30, 1970.

8. While the term "space station" is now universally used to describe a permanent, human-occupied laboratory in low-Earth orbit, this was not always the case. During initial concept studies, the NASA Langley Research Center called its version the MORL (Manned Orbiting Research Laboratory), which then became the MOWS (Manned Orbiting Workshop). The Manned Spacecraft Center named

its version MOM (Mission Operational Module), while the Marshall Space Flight Center had the MASL (Modular A Space Lab). Having all these names was not only confusing but, in time, became counterproductive, as each NASA field center pushed for its own design. Paine, then the acting administrator, told the Office of Manned Space Flight to abandon the use of disparate names and to just use the term "space station." (William J. Normyle, "NASA Molding Post-Apollo Plans," *Aviation Week and Space Technology*, December 23, 1968, 16–17.)

9. "Washington Roundup," *Aviation Week and Space Technology*, September 30, 1968, 15.

10. David W. Compton and Charles D. Benson, *Living and Working in Space: A History of Skylab* (Washington, DC: NASA, 1983), 104.

11. Skylab fell back to Earth and burned up on reentry July 11, 1979. Some large pieces were recovered in Western Australia.

12. "U.S./USSR Cooperation on Station Program Possible," *Space Daily*, May 11, 1970, 3; Paine, "Address to Australian Ministers and Officials."

13. "Civilian Space Stations and the U.S. Future in Space," report number OTA-STI-241, Washington, DC, Office of Technology Assessment, US Congress, November 1984, 166–168.

14. John M. Logsdon et al., *Managing the Moon Program: Lessons Learned From Apollo* (Washington, DC: NASA, 1999), ch. 8.

15. The original plan was to have DuBridge lead the STG.

16. Homer Newell did most of the daily legwork for Paine on the STG. U. Alexis Johnson represented the State Department, Glenn T. Seaborg the AEC, and Robert P. Mayo the Budget Bureau. (Hans M. Mark, *The Space Station: A Personal Journey* [Durham, NC: Duke University Press, 1987], 30–31; Paine, "Speech before the Air Force Association.")

17. Seamans, *Aiming at Targets,* 118.

18. Thomas O. Paine, interview with John M. Logsdon, September 3, 1970, transcript, 13–14.

19. Paine, "Address before the National Press Club."

20. "Proceedings of the Space Task Group Press Conference," proceedings of press conference, Office of the White House Press Secretary, September 17, 1969, 1–3.

21. Paine, Statement before the Senate Subcommittee on Independent Offices, Committee on Appropriations, May 19, 1970; "Presidential Task Force on Aeronautics Urged," Zack Strickland, *Aviation Week and Space Technology,*

December 15, 1969, 21; Zack Strickland, "NASA Reduces Manned Space Program," *Aviation Week and Space Technology,* September 7, 1970, 18–19.

22. Paine, Statement before the Senate Subcommittee on Independent Offices, Committee on Appropriations, May 19, 1970.

23. *Air Force and Space Digest*, December 1969, 37–44.

24. Paine, "Address to Australian Ministers and Officials."

25. Space shuttle flights from Vandenberg never took place, however, mostly due to the fallout from the *Challenger* accident in January of 1986.

26. Dennis R. Jenkins, *Space Shuttle: The History of the National Space Transportation System* (Stillwater, MN: Voyageur Press, 2001), 99.

27. "President Nixon's 1972 Announcement on the Space Shuttle," http://history.nasa.gov/stsnixon.htm.

28. NERVA stood for Nuclear Engine for Rocket Vehicle Application. It was a highly successful test program that never made it into the operational phase due to lack of funding. The program was canceled in 1972. Human space missions to Mars now being planned by NASA and private space companies for the decades of the 2020s and 2030s include the use and revival of advanced nuclear and ion, "mass driver" propulsion systems. Concepts for such systems were being developed in the 1950s and 1960s.

29. Thomas O. Paine, "Letter to the Vice President," Thomas Paine Associates, July 13, 1990.

30. Paine, Statement Before Senate Committee, March 1970; Logsdon et al., *Managing the Moon Program.*

31. William J. Normyle, "Manned Mission to Mars Opposed," *Aviation Week and Space Technology*, August 18, 1969, 16–17.

32. Paine, "Address before the National Press Club."

33. DeVoss, "So Says," 23–25.

34. Paine, interview by Logsdon and Cohen, 10–11; Mark, interview by author, 9; Philip J. Klass, "Commission Considers Joint Mars Exploration, Lunar Base Options," *Aviation Week and Space Technology*, July 29, 1985, 17.

11: I ACCEPT YOUR RESIGNATION

1. Reinhold, *New York Times,* November 15, 1969, 1.

2. Dolan, "Twentieth Century Version," 5; Paine, interview by Sherrod, September 10, 1970, 25.

3. Paine, interview by Emme, August 3, 1970, 27.

4. Paine, interview by Logsdon and Cohen, 19; Thomas O. Paine, "Memorandum for the Record: Meeting with the President," Office of the Administrator, NASA Headquarters, Washington, DC, January 22, 1970.

5. Paine, interview by Emme, August 3, 1970, 30–31; Paine, interview with Logsdon and Cohen, 21.

6. Paine, "Meeting with the President."

7. Thor Gawdiak, *NASA Historical Data Book*, vol. 4, *NASA Resources 1969–1978* (Washington, DC: National Aeronautics and Space Administration, 1994), ch. 4.

8. Ibid.

9. Paine, interview by Emme, August 3, 1970, 22–24.

10. Paine, interview by Emme, August 3, 1970, 30-31; Paine, interview by Logsdon and Cohen, 21.

11. Paine, interview by Sherrod, August 14, 1970, 36 – 37; Daniel L. Dumbacher, comments to author, May 6, 2016.

12. Washington Roundup, *Aviation Week and Space Technology*, May 18, 1970, 13.

13. Yuri Gagarin, the first person in space, had been killed in a training accident in 1968. In 1998, John Glenn flew into space again as a Payload Specialist on STS-95. At age 77, he was the oldest person to go into space.

14. Armstrong, e-mail to author, December 3, 2011.

15. Paine, interview by Sherrod, October 7, 1971, 4–5.

16. Neil Armstrong, e-mail to author, April 2, 2011.

17. Paine, interview by Sherrod, May 13, 1969, 1.

18. NASA News Release 69-151, NASA Headquarters, Washington, DC, November 10, 1969.

19. Swigert had replaced Ken Mattingly as command module pilot less than seventy-two hours before launch when it was discovered that Mattingly might have been exposed to the German measles. The crew change-out was questioned and debated up until the day before launch. At the T-24 hour Flight Readiness Review, Paine signed off on the decision to launch after meeting with the Apollo management team at KSC and after having conferred privately with Mission Commander Lovell and Lunar Module Pilot Haise.

20. John Lannan, "Paine Sees Setback to Program," *Washington Star*, April 15, 1970, A1.

21. Erwin J. Bulban, "Apollo 13 Crises Spur Massive Support," *Aviation Week and Space Technology,* April 27, 1970, 22–27.

22. Blue, "Houston."

23. Thomas O. Paine, "Statement for the Apollo 13 Review Board Press Conference," NASA Headquarters, Washington, DC, June 15, 1970.

24. Paine, "Statement before the Senate Committee on Aeronautical and Space Sciences," June 30, 1970.

25. John Lannan, "NASA Chief Seeks to Save Space Goals," *Washington Star,* April 24, 1970, A1.

26. Administrator Chronological Files, July 1970.

27. Julian Scheer was not in the private meeting in which Paine tendered his resignation.

28. "Exchange of Letters Between the President and Dr. Thomas O. Paine on Dr. Paine's Resignation as Administrator of NASA, 28 July 1970," from the Weekly Compilation of Presidential Documents, the White House, vol. 6, no. 31, p. 997, August 3, 1970.

29. Victory K. McElheny, "NASA Officials at Woods Hole Surprised by Paine Resignation," *Boston Globe,* July 30, 1970, 5–6; Mark, interview by author, 11–12.

30. Thomas O. Paine, "Letter to NASA Employees from Administrator Thomas O. Paine," NASA Headquarters, Washington, DC, July 28, 1970.

31. Dr. Thomas O. Paine's Resignation as NASA Administrator, transcript of news conference, NASA Headquarters, Washington, DC, July 29, 1970.

32. "NASA Chief Analyzes U.S. Space Plan," *Chicago Tribune,* July 5, 1970, 1.

33. Paine, interview by Emme, August 3, 1970, 21–22; Dumbacher, comments to author.

34. Paine, interview by Sherrod, August 14, 1970, 28.

35. Ibid., 13–15.

36. Administrator Chronological Files, September 1970.

37. Mark, interview by author, 16.

38. Thomas O. Paine, "Remarks at Farewell Dinner," speech, Bolling Air Force Base, Washington, DC, September 15, 1970.

39. Thomas O. Paine, "Apollo and Mars," *Ad Astra,* July/August 1989, 3.

40. Paine, interview by Emme, September 3, 1970, 11.

41. Paine, "Apollo and Mars"; Dolan, "Twentieth Century Version," 6; Paine, interview by Emme, September 3, 1970, 21.

42. Buckley, "NASA's Tom Paine," 50.

43. Paine, interview by Emme, September 3, 1970, 31–32.

44. Thomas O. Paine, "Statement by Administrator," transcript of proceedings, NASA Headquarters, Washington, DC, February 27, 1969.

45. Thomas O. Paine, "Address at the Alfred E. Smith Dinner," speech, New York, October 16, 1969.

12: A LITTLE BETTER FOOTING

1. Paine, interview by Sherrod, October 7, 1971, 25–26.

2. Paine, interview by Emme, August 3, 1970, 12–14.

3. Thomas O. Paine, "Outline of Factors Relevant to National Energy Policy Formulation," draft report, General Electric, January 9, 1973; "Fuel of the Future," *San Francisco Chronicle,* April 30, 1974.

4. Paine, "Outline of Factors."

5. P. C. Luttenberger, "Letter to C. Marchetti and R. Schulten," internal memorandum, GE TEMPO, n.d.

6. Thomas O. Paine, "Factors Involved in National Energy Policies," speech to the Morgan Guaranty International Council, San Francisco, California, February 9, 1973.

7. Thomas O. Paine, "Letter to the PGBG Executive Board," internal memorandum, General Electric, March 6, 1973.

8. Peter M. Flanigan, "Letter to Tom Paine," personal correspondence from the White House, January 26, 1973; Paine, "Factors Involved"; Paine, "Morgan Guaranty."

9. Reginald H. Jones replaced Fred J. Borch as the CEO of General Electric in 1972.

10. The State Committee for Science and Technology was, at the time, the authoritative body responsible for all industrial research and development in the Soviet Union.

11. Theodore Shabad, "G.E. and Soviet Sign Pact for Technology Exchange," *New York Times,* January 13, 1973, 1.

12. *The Oregonian* and *San Jose Mercury,* January 13, 1973, 1.

13. Thomas H. Lee, "Letter to Paine," correspondence 8-245-3220, GE Power Delivery Group, Philadelphia, PA, April 17, 1972; "Itinerary for Dr. Thomas O. Paine," internal memorandum, General Electric, June 1972.

14. Yen Chia Kan, "Note to Dr. Thomas O. Paine," letter from the Office of the Vice President, Republic of China, July 1972.

15. W. Ridgway, "Letter to C. K. Yen," correspondence from General Electric Headquarters to the Vice President of the Republic of China, August 9, 1972. The author's father, Tsiao Tso Fan, was a trainee from the Republic of China in 1968.

16. Thomas O. Paine, "Letter to Reginald Jones," handwritten draft memorandum for the CEO of General Electric, 1975, 1–8.

17. Ibid., 18–22.

18. Ibid., 35, 51.

19. Edward E. Hood Jr., interview by author, June 14, 2011, transcript, 5. The former General Electric executive recalled that Paine's resignation was, at the time, "a non-event." The power transfer atop GE from Reg Jones to a handpicked successor has been analyzed by corporate business insiders for years and is the topic of numerous books. One notable volume is the autobiography *Jack: Straight from the Gut*, written by John F. Welch Jr., who took over the company in 1981 and completely transformed GE over the next two decades into the multinational conglomerate that it is today.

20. Thomas O. Paine, "1975 Resume," Corporate Files, LOC, 1975.

21. T. J. Johnston, "Letter to Paine," correspondence from Heidrick and Struggles, Inc., to Paine, November 26, 1975.

22. The Northrop Corporation acquired Grumman Aerospace in 1994 to form Northrop Grumman. In 2016 it was the eighth largest aerospace company in the world, with revenues of over $23 billion. ("The World's 20 Largest Aerospace and Defense Companies as Measured by 2015 Revenue," *CEOWORLD Magazine*, July 2015, ceoworld.biz/2016/07/15/worlds-20-largest-aerospace-defense-companies-measured-2015-revenue.)

23. Thomas O. Paine, "Notes on Executive Office," handwritten notes on Northrop management spectrum, Corporate Files, LOC, n.d.

13: PIONEERING THE SPACE FRONTIER

1. *Aviation Week and Space Technology*, February 5, 1979, 12.

2. Peter Cobon, "Ex-Directors Blast Carter's Space Effort as Lip-Service," *Huntsville Times*, July 24, 1980, 6.

3. James E. Webb, "Letter to Dr. Thomas O. Paine," personal correspondence, January 23, 1981; Karl G. Harr Jr., "Letter to Thomas O. Paine, President and COO Northrop Corporation," personal correspondence, January 28, 1981.

4. Mark, interview by author, 19.

5. Jay Keyworth, interview by author, February 2, 2011, 5.

6. Thomas O. Paine, "Letter to Dr. George M. Low," personal correspondence, January 14, 1981.

7. Harr, "Letter to Paine."

8. Beggs, interview by Rusnak, 19–21.

9. Thomas O. Paine, "Letter to Dr. J. E. Pournelle," personal correspondence, January 23, 1981.

10. Keyworth, interview by author, 4.

11. Executive Order Number 12490 as authorized by the US Congress under PL 98-361.

12. "Reagan Presses Call for Antimissile Plan before Space Group," *New York Times*, March 30, 1985, 1, 41.

13. Keyworth, interview by author, 2–3.

14. National Commission on Space (NCOS), appendix to *Pioneering the Space Frontier* (New York: Bantam, 1986).

15. NCOS, "First Meeting" agenda, May 15–17, 1985.

16. Thomas Paine Associates, "Notes on memorandum to the NCOS commissioners," June 8, 1985.

17. Lee Dye, "Charting the Future in Space," *Los Angeles Times*, July 19, 1985, 5; Keyworth, interview by author, 7.

18. Laurel L. Wilkening, interview by author, December 1, 2010, transcript, 6–7; Wilkening, interview by Butler, 61.

19. NCOS, "NCOS Public Forum Strategy, August 21, 1985.

20. *Meet the Press*, proceedings of the television program, National Broadcasting Company, New York, vol. 86, February 2, 1986.

21. Armstrong, e-mail to author, March 12, 2011.

22. NCOS, "First Meeting."

23. Wilkening, interview by Butler, 54–60.

24. Ed Rogers, "Manned Flights Seen Crucial to Space Program's Future," *Washington Times*, June 9, 1986, A3.

25. NCOS, preface to *Pioneering the Space Frontier.*

26. NCOS, *Pioneering the Space Frontier*, ch. 2.

27. Armstrong, e-mail to author, March 12, 2011; Wilkening, interview by author, 8.

28. In July 2011, the Space Shuttle program ended with the landing of STS-135 at the Kennedy Space Center. From 1981 to 2011, the shuttle flew 135 times. There were five orbiter vehicles: *Columbia, Challenger, Discovery, Atlantis, and Endeavour. Challenger* was lost during launch in 1986, and *Columbia* was lost on reentry in 2003.

29. "Commission May Suggest More Space Station," *Aviation Week and Space Technology*, October 14, 1985, 18.

30. A spacecraft in geosynchronous orbit at approximately 22,000 miles above mean sea level revolves around Earth at the same rate as a location on the surface of the Equator. It would thus appear, to an observer on the ground, to not move over that location. Geosynchronous orbit (also known as geostationary orbit) is thus ideal for communication and weather satellites.

31. NCOS, preface to *Pioneering the Space Frontier.*

32. DeVoss, "So Says," 23–25.

33. Donald E. Fink, *Aviation Week and Space Technology*, March 24, 1986, 13.

34. Michael Isikoff, "NASA Chief Hails Push for Mars Colony," *Washington Post*, May 24, 1986, A3.

35. Ibid.

36. Keyworth, interview by author, 11.

37. Reagan had presented L'Amour with the Medal of Freedom in 1984.

38. Wilkening, interview by author, 5; Wilkening, interview by Butler, 54.

39. Armstrong, e-mail to author, March 8, 2011.

14: CHIEF MARTIAN MONSTER

1. Frutkin, interview by Wright, 53–56.

2. The House Committee members on the trip were: Clarence W. (Bill) Nelson II, Chairman of the House Subcommittee on Space Science and Applications; Robert L. Coughlin; Manuel Lujan, Jr.; Donald L. Ritter; Robert G. Torricelli; Robert S. Walker; and Robert A. Young. The third member of the ASTP crew, Vance Brand, had commitments as an active shuttle astronaut and could not attend.

3. Thomas O. Paine, "Addendum to impressions received in Moscow relating to the future direction of the Soviet space program," personal memorandum to NASA Administrator James Beggs, October 26, 1985.

4. Ibid.

5. What was left of *Buran*, along with other rockets and ground-support equipment, was destroyed when the hangar collapsed during a massive storm in 2002.

6. Thomas O. Paine, "Observations on a trip to Moscow to attend the International Space Future Forum celebrating the 30th anniversary of Sputnik," personal memorandum to NASA Administrator James Fletcher, October 19, 1987.

7. Executive Order 12675.

8. Dwayne A. Day, "A New Space Council?" *Space Review*, June 21, 2004, http://www.thespacereview.com/article/163.

9. "20 Years after Apollo: Is the U.S. Lost in Space?" *Popular Science*, July 1989.

10. David S. F. Portree, *Humans to Mars: Fifty Years of Mission Planning 1950–2000* (Washington, DC: NASA, 2001), 83.

11. John N. Wilford, "Bush Sets Target for Mars Landing," *New York Times*, May 12, 1990, 9.

12. Paine, "Apollo and Mars," 3.

13. Wilkening, interview by author, 15.

14. Paine, "Letter to the Vice President."

15. Freidman, interview by author, March 8, 2011, 9–10.

16. "Space Transportation Costs: Trends in Price Per Pound to Orbit 1990–2000," technical report of the Futron Corporation, Bethesda, Maryland, September 6, 2002.

17. Paine, "Letter to the Vice President"; Sagan et al., "The Last of the Corsairs," 6.

18. "20 Years after Apollo;" Paine, "Letter to the Vice President."

19. Thor Hogan, *Mars Wars: The Rise and Fall of the Space Exploration Initiative* (Washington, DC: NASA, 2007), 98–100.

20. Paine, "The Space Commission Lives!" memorandum to the members of the NCOS, September 4, 1990.

21. Paine, fax to James D. Bain, August 29, 1990.

22. Wilkening, interview by author, 17.

23. Norman R. Augustine et al., "Principal Recommendations and Actions Taken, Advisory Committee on the Future of the U.S. Space Program,"

presentation of the ACFUSSP to the National Space Council, Washington, DC, September 12, 1991.

24. Paine, "Letter to Dr. Laurel Wilkening," personal correspondence, October 11, 1990.

25. Wilkening, interview by author, 9.

26. Paine, "A note to Norm Augustine," handwritten note filed in the General Correspondence Files, LOC, October 20, 1990.

27. Paine, foreword to a draft outline of the ACFUSSP report, October 27, 1990.

28. "Final Report of the ACFUSSP," 48–49; Augustine et al., "Principal Recommendations."

29. Marsha Freeman, "Crazy Panel Urges End to Programs for Man in Space," *New Federalist*, December 14, 1990, 1, 9–10; Paine, fax to Wilkening, January 3, 1990.

EPILOGUE: A TWINKLE IN HIS EYE

1. Thomas O'Toole, "Owning a Piece of the Rock," *Washington Post*, March 1, 1974, C1.

2. Thomas O. Paine, "A note to Gene Emme," November 8, 1973.

3. Paine, "Letter to The President and Letter to Dr. James C. Fletcher," personal correspondence, November 2, 1973.

4. James C. Fletcher, "Letter to Dr. Thomas O. Paine," personal correspondence, February 5, 1974.

5. Richard T. Feller, *A Cathedral Window Commemorating the Exploration of Space*, information brochure by the Clerk of the Works, Washington Cathedral, Washington, DC, n.d. The mineral pyroxferroite, unknown on Earth, was discovered in this rock brought back on Apollo 11. The fragment itself was kept sealed in a vault and not actually installed into the window until one and a half years after the dedication service.

6. Charles Martin, "Letter to Dr. Thomas O.Payne [*sic*]," personal correspondence, St. Albans School, Washington, DC, January 26, 1974. Apollo 11 Command Module Pilot Mike Collins was, incidentally, an alumnus of St. Albans.

7. Paine, "Letter to The President recommending construction of the National Air and Space Museum," correspondence to Richard Nixon, n.d.

8. Eugene M. Emme, “The Logical Apollo 11 Memorial: Construction of the National Air and Space Museum on the Mall,” correspondence to the Administrator of NASA, August 13, 1969.

9. Friedman, interview by author, 2.

10. Ibid., 7; “Our Mission,” accessed May 7, 2017, http://www.planetary.org/about.

11. Friedman interview, 8.

12. Thomas O. Paine, “Humanity Unlimited,” *Newsweek*, August 25, 1975, 3.

13. Thomas O. Paine, “Man’s Future in Space,” Tizard Memorial Lecture, Westminster School, London, March 14, 1972.

14. Thomas O. Paine, “On the Year 2000,” speech at the Young Presidents’ Organization, St. George, Bermuda, October 10, 1968.

15. Thomas O. Paine, “SMS Vulcan and HIJMS I-400,” technical memorandum to the SWATH Task Force, TPA, Los Angeles, CA, April 27, 1984. On the engineering front, he was a technical consultant for the Navy on the SWATH (Small Waterplane Area Twin-Hull) Task Force, doing research and development of twin-hull ocean surveillance vessels—massive ships that today conduct surveillance and long duration, deep-sea missions for the US Navy.

16. Paine, “Sharply etched.”

17. “Nimitz Library in the News,” press release, US Naval Academy, Annapolis, MD, http://www.usna.edu/library/news/libnews.html.

18. “Mars Observer to Carry Tribute to NASA’s Paine,” *Space News*, August 31–September 6, 1992, 29.

19. Keyworth, interview by author, 9.

20. Blue, “Houston.”

BIBLIOGRAPHY

INTERVIEWS

Armstrong, Neil A., by Sunny Tsiao, via e-mail, February–December 2011.

Beggs, James M., by Kevin Rusnak, March 7, 2002.

Cortright, Edgar M., by Rich Dinkel, August 20, 1998.

DuBridge, Lee A., by Judith R. Goodstein, February 19–20, 1981.

Finger, Harold B., by Kevin Rusnak, May 16, 2002.

Franklin, George C., by Kevin Rusnak, Sandra Johnson, and Jennifer Ross-Nazzal, October 3, 2001.

Friedman, Louis D., by Sunny Tsiao, March 8, 2011.

Frutkin, Arnold W., by Rebecca Wright, March 8, 2002.

Hood, Edward E., Jr., by Sunny Tsiao, June 14, 2011.

Johnson, Lyndon B., by Robert Sherrod, July 25, 1972.

Jones, Reginald H., by Gerald Zahavi, June 12, 2000.

Keyworth, George A., II, by Sunny Tsiao, February 2, 2011.

Kraft, Christopher C., Jr., by Rebecca Wright, Sandra Johnson, and Jennifer Ross-Nazzal, May 23, 2008.

Lovelace, Alan M., by Sandra Johnson and Rebecca Wright, July 14, 2011.

Low, George M., by John Logsdon, July 7, 1970.

Low, George M., by Robert Sherrod, August 12, 1970.

Mark, Hans M., by Sunny Tsiao, January 29, 2011.

Mueller, George E., by Robert Sherrod, April 21, 1971.

Mueller, George E., by Summer Chick Bergen and Carol Butler, August 27, 1998.

Myers, Dale D., by Carol Butler, Summer Chick Bergen, and Rebecca Wright, August 26, 1998 and March 5, 1999.

Paine, Thomas O., by Arthur Garratt, June 9, 1975.

Paine, Thomas O., by Eugene M. Emme, July 9, 1970; August 3, 1970; and September 3, 1970.

Paine, Thomas O., by John M. Logsdon, September 3, 1970.

Paine, Thomas O., by John M. Logsdon and Nat Cohen, August 12, 1970.

Paine, Thomas O., by Larry Suid, n.d.

Paine, Thomas O., by Robert Sherrod, November 5, 1968; January 23, 1969; May 13, 1969; October 13, 1969; August 14, 1970; September 10, 1970; and October 7, 1971.

Paine, Thomas O., by T. Harri Baker, March 25, 1969 and April 10, 1969.

Paine, Thomas O., by V. Kobysh, June 16, 1973.

Paine, Thomas O., *Issues and Answers,* ABC, March 15, 1970.

Paine, Thomas O., *Meet the Press* , NBC, May 18, 1969.

Paine, Thomas O., *Meet the Press,* NBC, November 16, 1969.

Paine, Thomas O., *Meet the Press,* NBC, February 2, 1986.

Seamans, Robert C., Jr., by Carol Butler and Rebecca Wright, June 22, 1999.

Seamans, Robert C., Jr., by Michelle Kelly, September 30 and November 20, 1998.

Scheer, Julian W., by Robert Sherrod, June 29, 1970 and February 19, 1971.

Soper, Josie A., by Rebecca Wright, April 19, 2006.

Webb, James E., by H. George Frederickson, May 15, 1969.

Webb, James E., by Robert Sherrod, October 7, 1971.

Webb, James E., by T. Harri Baker, April 29, 1969.

Wilkening, Laurel L., by Carol Butler, November 15, 2001.

Wilkening, Laurel L., by Sunny Tsiao, December 1, 2010.

SPEECHES BY THOMAS O. PAINE

"The City as an Information Network." Presentation at the Institute of Electrical and Electronics Engineers International Convention, New York, NY, March 22, 1966.

"The City of the Future." Presentation at the Architects and Engineers Forum, Los Angeles, CA, April 27, 1968.

Untitled presentation at the NASA Management Advisory Panel, Washington, DC, May 14, 1968.

"On the Year 2000." Presentation at the Young Presidents' Organization, St. George, Bermuda, October 10, 1968.

"1969: A Space Odyssey." Presentation at the American Institute of Aeronautics and Astronautics, Washington, DC, November 7, 1968.

Untitled presentation at the Aerospace Industries Association, Phoenix, AZ, November 20, 1968.

Untitled presentation at the Franklin Institute, Philadelphia, PA, March 4, 1969.

Thomas A. Edison Memorial Lecture, Washington, DC, March 11, 1969.

Untitled presentation at the Air Force Association, Houston, TX, March 20, 1969.

"The Relevance to Cities of Space Age Systems Management." Presentation at the National Conference on Public Administration, Miami Beach, FL, May 20, 1969.

Untitled presentation at the Commonwealth Club, San Francisco, CA, August 1, 1969.

Remarks to the National Press Club, Washington, DC, August 6, 1969.

Remarks at the Apollo 11 Presidential State Dinner, Los Angeles, CA, August 13, 1969.

Address to the United Nations Committee on the Peaceful Uses of Outer Space, New York, NY, September 8, 1969.

Untitled presentation at the Clarkson College of Technology, Potsdam, NY, September 26, 1969.

Remarks at the Alfred E. Smith Foundation dinner, New York, NY, October 16, 1969.

Untitled presentation at the American Association for the Advancement of Science, Boston, MA, December 26, 1969.

Untitled presentation at the annual NASA industry briefing, Washington, DC, February 6, 1970.

Remarks to the Morgan Guaranty International Council, San Francisco, CA, February 9, 1970.

Untitled presentation to Australian ministers and officials, Canberra, Australia, February 26, 1970.

Commencement address at Nebraska Wesleyan University, Lincoln, NE, May 2, 1970.

"The Space Age." Commencement address at the University of New Brunswick, St. John, Canada, May 14, 1970.

Remarks at the Apollo History Workshop, Washington, DC, May 21, 1970.

Untitled speech to the European Space Research Organization, Paris, France, June 3, 1970.

"Squareland, Potland, and Space." Commencement address at Worcester Polytechnic Institute, Worcester, MA, June 7, 1970.

Remarks at the NASA Long-Range Planning Conference, Washington, DC, June 14, 1970.

Presentation of moon rock to the United Nations, New York, NY, July 20, 1970.

"What Lies Ahead in Space." Presentation to the Economic Club of Detroit, Detroit, MI, September 14, 1970.

Remarks at his NASA farewell dinner, Bolling Air Force Base, Washington, DC, September 15, 1970.

Remarks at his NASA portrait unveiling ceremony, Washington, DC, November 9, 1971.

"Man's Future in Space." Tizard Memorial Lecture to the Westminster School, London, UK, March 14, 1972.

"Factors Involved in National Energy Policies." Presentation to the Morgan Guaranty International Council, San Francisco, CA, February 9, 1973.

Commencement address at the University of Hawaii, Honolulu, HI, May 17, 1981.

"Leaving the Cradle: Human Exploration of Space in the 21st Century." Welcome and concluding remarks from the proceedings of the 28th Goddard Memorial Symposium, AAS 90-101, 90-129, American Astronautical Society, Washington, D.C., August 1991.

PRESS CONFERENCES WITH THOMAS O. PAINE

Apollo 7 mission, Washington, DC, September 23, 1968.

Apollo 8 mission, Washington, DC, November 12, 1968.

FY 1970 NASA budget, Washington, DC, January 15, 1969.

Apollo 9 mission, Washington, DC, February 27, 1969.

FY 1970 NASA budget, Washington, DC, April 15, 1969.

Apollo 11 mission, Houston, TX, July 20, 1969.

Closing of the Electronics Research Center, Cambridge, MA, December 29, 1969.

NASA future plans, Washington, DC, January 13, 1970.

FY 1971 NASA budget, Washington, DC, February 2, 1970.

President Nixon's press conference on the future of the US space program, Washington, DC, March 7, 1970.

Apollo 13 mission, Washington, DC, April 19, 1970.

Launch of the first Chinese satellite, Washington, DC, April 25, 1970.

Findings of the Apollo 13 Accident Review Board, Washington, DC, June 15, 1970.

Resignation as NASA Administrator, San Clemente, CA, July 29, 1970.

1971 NASA interim operating plan, Washington, DC, September 2, 1970.

White House press conference on the National Commission on Space, Washington, DC, March 29, 1985.

CONGRESSIONAL TESTIMONIES OF THOMAS O. PAINE

Statement before the House Subcommittee on Advanced Research and Technology, September 24, 1968.

Statement before the Senate Committee on Aeronautical and Space Sciences, October 3, 1968.

Statement before the House Committee on Science and Astronautics, March 4, 1969.

Confirmation hearing before the Senate Committee on Aeronautical and Space Sciences, March 14, 1969.

Statement before the House Subcommittee on Appropriations, April 15, 1969.

Statement before the Senate Subcommittee on Government Research, April 22, 1969.

Statement before the Senate Committee on Aeronautical and Space Sciences, April 28, 1969.

Statement before the Senate Subcommittee on Appropriations, July 10, 1969.

Statement before the House Committee on Science and Astronautics, October 1, 1969.

Statement before the Senate Committee on Aeronautical and Space Sciences, February 2, 1970.

Statement before the House Committee on Science and Astronautics, February 17, 1970.

Statement before the Senate Committee on Aeronautical and Space Sciences, February 20, 1970.

Statement before the Senate Committee on Aeronautical and Space Sciences, March 11, 1970.

Statement before the House Committee on Appropriations, Independent Offices Subcommittee, March 24, 1970.

Statement before the Senate Committee on Aeronautical and Space Sciences, April 6, 1970.

Statement before the Senate Committee on Aeronautical and Space Sciences, April 24, 1970.

Statement before the Senate Committee on Appropriations, Subcommittee on Independent Offices, May 19, 1970.

Statement before the Senate Committee on Appropriations, Subcommittee on Deficiencies and Supplementals, May 28, 1970.

Statement before the Senate Committee on Aeronautical and Space Sciences, June 30, 1970.

Statement before the Senate Subcommittee on Science, Technology, and Space, July 22, 1986.

WRITINGS OF THOMAS O. PAINE

"Southwest Pacific 1943." Wartime journal, July 1945–January 1946.

"The Effect of a Molten Bismuth Eutectic Alloy on Steel." PhD diss., Stanford University, 1949.

"Effect of Shape Anisotropy on the Coercive Force of Elongated Single-Magnetic-Domain Iron Particles." *Physical Review* 100, November 1955.

"Reproducing the Properties of Alnico Permanent Magnet Alloys with Elongated Single-Domain Cobalt-Iron Particles." *Journal of Applied Physics* 28, no. 3, March 1957.

"Coercive Force and Remanence of 25-A to 2000-A Diameter Cobalt, Iron, and Iron-Cobalt Alloy." *Journal of Applied Physics* 31, no. 5, May 1960.

"The City as an Information Network." General Electric Tempo report no. 66TMP32, Santa Barbara, CA, March 22, 1966.

"Interim Report on Urban Development Corporation." Draft report, General Electric Tempo, Santa Barbara, CA, September 2, 1966.

"Concept of a Space Age City." Paper for the Architects and Engineers Forum, Los Angeles, CA, April 27, 1968.

"Cities: Who Can Save Them?" *Look*, June 11, 1968.

"Space Age Administration." *Ordnance* 54, no. 296, 1969.

"Space Research and a Better Earth: Impact of Science on Society." *Impact of Science on Society* (UNESCO) 19, no. 2, 1969.

"Space and National Security in the Modern World." *Air Force and Space Digest*, May 1969.

Untitled story on Apollo 11, *Die Welt* magazine, July 1969.

"Space and Man's Future." *Pace* 5, no. 8, August 1969.

"Report on the Flight of Apollo 11." Report to the United Nations Committee on the Peaceful Uses of Outer Space, USUN-94, September 8, 1969.

"Space Age Management and City Administration." *Public Administration Review* November/December 1969.

"Next Steps in Space." *National Geographic* 136, no. 6, December 1969.

"Congressional Recognition of Goddard Rocket and Space Museum with Tributes to Dr. Robert H. Goddard." Committee on Aeronautical and Space Sciences, United States Senate document no. 91-71, Washington, DC, May 28, 1970.

"Submarining: Three-Thousand Books and Articles." General Electric TEMPO Center for Advanced Studies, Santa Barbara, CA, 1971.

"Man's Future in Space." *Contemporary Physics* 13, issue 4, July 1972.

"Outline of Factors Relevant to National Energy Policy Formulation." Draft report, General Electric, New York, NY, January 9, 1973.

"Humanity Unlimited." *Newsweek*, August 25, 1975.

"The Global Implication of Space." Unpublished paper, Los Angeles, CA, August 1981.

"Space: The Future of Mankind." Unpublished paper, Santa Barbara, CA, 1983.

"The Transpacific Voyage of His Imperial Japanese Majesty's Submarine I-400." Draft of personal narrative of wartime journal, Los Angeles, CA, 1984.

"A Timeline for Martian Pioneers." American Astronautical Society paper AAS84-150, 1984.

"SMS Vulcan and HIJMS I-400." Memorandum to SWATH Task Force, Thomas Paine Associates, Santa Barbara, CA, April 27, 1984.

"Wrong Since Goddard." Written with Arthur C. Clarke and George E. Mueller. Letter to the editor, *New York Times,* December 7, 1984.

"I Was a Yank on a Japanese Sub." Naval Institute Press, U.S. Navy, vol. 112/9/1003, Annapolis, MD, September 1986.

"Apollo and Mars." *Ad Astra,* National Space Society, July/August 1989.

"The Rationale for Mars and Lessons from Apollo." American Astronautical Society paper AAS90-318, June 1990.

"Sharply Etched Unforgettable Images from the World War II Pacific Submarine War." Personal notes, April 1992.

"Mars Colonization: Technically Feasible, Affordable, and a Universal Human Drive," Strategies for Mars: A Guide to Human Exploration, American Astronautical Society paper AAS95-495, published posthumously, 1996.

SELECTED BIBLIOGRAPHY

Bartlett, John R. "Records of the Colony of Rhode Island and Providence Plantations in New England," vol. 2, 564. https://archive.org/search.php?query=creator%3A%22Bartlett%2C+John+Russell%2C+1805-1886%22.

Brooks, Courtney G., James M. Grimwood, and Loyd S. Swenson, Jr. *Chariots for Apollo: The NASA History of Manned Lunar Spacecraft to 1969.* Mineola, NY: Dover Publications, 2009.

Buckley, Thomas. "NASA's Tom Paine: Is This a Job for a Prudent Man?" *New York Times Magazine*, June 1969.

Committee on the Future of the US Space Program. "Report of the Advisory Committee on the Future of the U.S. Space Program." Presidential Committee Report, Washington, DC, December 17, 1990.

Compton, W. David, and Charles D. Benson. *Living and Working in Space: A History of Skylab.* Washington, DC: National Aeronautics and Space Administration, 1983.

———. *Where No Man Has Gone Before: A History of Apollo Lunar Exploration Missions.* Washington, DC: National Aeronautics and Space Administration, 1989.

Cortright, Edgar M. *Apollo Expeditions to the Moon: The NASA History.* Mineola, NY: Dover Publications, 2009.

David, Edward E. "Advice for President Richard Nixon." Presentation at the University of Colorado, Boulder, CO, September 12, 2005.

DeVoss, David. "So Says Thomas O. Paine." *Los Angeles Times Magazine*, April 13, 1986.

Dick, Steven J. *NASA's First 50 Years: Historical Perspectives.* Washington, DC: National Aeronautics and Space Administration, 2010.

Dolan, Patricia N. "Thomas Paine, Twentieth Century Version." *Challenge*, General Electric Company, Fall 1969.

Eisenhower, Julie Nixon. *Pat Nixon: The Untold Story.* New York: Simon and Schuster, 1986.

Ezell, Edward C., and Linda Neuman Ezell. *The Partnership: A History of the Apollo-Soyuz Test Project.* Washington, DC: National Aeronautics and Space Administration, 1978.

Ezell, Linda Neuman. *NASA Historical Databook.* Vol. 3, *Programs and Projects 1969–1978.* Washington, DC: National Aeronautics and Space Administration, 1994.

Feller, Richard T. "A Cathedral Window Commemorating the Exploration of Space." Report of the Clerk of the Works, Washington Cathedral, Washington, DC, n.d.

Futron Corporation. "Space Transportation Costs: Trends in Price per Pound to Orbit 1990–2000." Contractor report. Bethesda, MD, September 6, 2002.

Gawdiak, Thor. *NASA Historical Data Book*. Vol. 4, *NASA Resources 1969–1978*. Washington, DC: National Aeronautics and Space Administration, 1994.

Hansen, James R. *First Man: The Life of Neil A. Armstrong*. New York: Simon and Schuster, 2005.

Heppenheimer, T. A. *The Space Shuttle Decision*. Washington, DC: National Aeronautics and Space Administration, 1999.

Hogan, Thor. *Mars Wars: The Rise and Fall of the Space Exploration Initiative*. Washington, DC: National Aeronautics and Space Administration, 2007.

Hood, Edward E., Jr. "Thomas O. Paine." Memorial tribute of the National Academy of Engineering, vol. 6, Washington, DC, 1993.

Jenkins, Dennis R. *Space Shuttle: The History of the National Space Transportation System*. Stillwater, MN: Voyageur Press, 2001.

Kraft, Christopher. *Flight: My Life in Mission Control*. New York: Penguin Putnam, 2001.

Lambright, W. Henry. *Powering Apollo: James E. Webb of NASA*. Baltimore: Johns Hopkins University Press, 1998.

Lawton, A. T. "Dr. Thomas Paine: An Appreciation." *Spaceflight* 34, September 1992.

Levine, Arnold S. *Managing NASA in the Apollo Era*, Washington, DC: National Aeronautics and Space Administration, 1982.

Logsdon, John M. *After Apollo? Richard Nixon and the American Space Program*. Basingstoke, UK: Palgrave Macmillan, 2015.

———. *Managing the Moon Program: Lessons Learned From Apollo*. Washington, DC: National Aeronautics and Space Administration, 1999.

Mark, Hans M. "A Cold War Memoir." Chapter 6 of unpublished memoir. College Station, TX, September 2007.

———. *The Space Station: A Personal Journey*. Durham, NC: Duke University Press, 1987.

National Commission on Space. *Pioneering the Space Frontier: The Report of the National Commission on Space*. New York: Bantam Books, 1986.

Newkirk, Roland W., Ivan D. Ertel, and Courtney G. Brooks. *Skylab: A Chronology*. Washington, DC: National Aeronautics and Space Administration, 1977.

Office of Technology Assessment. "Civilian Space Stations and the U.S. Future in Space: OTA-STI-241." Congress of the United States, Washington, DC, November 1984.

O'Neill, Gerard K. *The High Frontier: Human Colonies in Space.* New York: William Morrow, 1976.

Peterson, William H. "Boulwarism: Ideas Have Consequences." *Freeman* 41, issue 4, April 1991.

Platoff, Anne M. "Where No Flag Has Gone Before: Political and Technical Aspects of Placing a Flag on the Moon." NASA Contractor Report 188251. Houston, TX, October 11, 1992.

Portree, David S. F. *Humans to Mars: Fifty Years of Mission Planning, 1950–2000.* Washington, DC: National Aeronautics and Space Administration, 2001.

Sagan, Carl E., Bruce C. Murray, Louis D. Friedman, and Charlene M. Anderson. "The Last of the Corsairs: The Life of Thomas O. Paine." *Planetary Report* 12, no. 5, Pasadena, CA, September/October 1992.

Seamans, Robert C., Jr. *Aiming at Targets.* Washington, DC: National Aeronautics and Space Administration, 1996.

———. *Project Apollo: The Tough Decisions.* Washington, DC: National Aeronautics and Space Administration, 2005.

Simpson, T. R. "A Review of The Space Station: An Idea Whose Time Has Come." Institute of Electrical and Electronics Engineers. *Spectrum* 22, no. 9, September 1985.

Smith, Marcia S. "Don't Forget the National Commission on Space (NCOS) Report!" www.SpacePolicyOnline.com, July 14, 2009.

"Space Task Group Report." Proceedings of the White House Press Conference, Office of the White House Press Secretary, Washington, DC, September 17, 1969.

Swanson, Glen E. *Before This Decade Is Out.* Washington, DC: National Aeronautics and Space Administration, 1999.

Zimmerman, Robert. *Genesis: The Story of Apollo 8.* New York: Random House, 1998.

INDEX

Abernathy, Ralph, 93–95
Advisory Committee on the Future of the US Space Program, 194–198
Agnew, Spiro, 73, 78, 84, 104, 124–125, 129, 140
Aiming at Targets, 47
Air Force and Space Digest, 126
Air Force One, 142
Air Force Two, 103, 194
Aldrin, Buzz, xii, xiii, 91, 92, 166, 202
  celebrations of, 103
  moon landing, 96–98, 101
Allen, Joseph P., 80–81
Alvarez, Luis W., 171
American President Lines, 3
American Revolution, 1–2
Ames Research Center, 147
Anders, Bill, 72–73
Anderson, Clinton P., 82, 99, 150
Apollo Project, xii, xvi. *see also* NASA
  Apollo 6, 150
  Apollo 7, 61–69, 86
  Apollo 8, 68–74, 83–84, 86, 150
  Apollo 9, 84, 86–87
  Apollo 10, 86–87, 90
  Apollo 11, 85, 90, 95–100, 108–109, 114, 134, 136, 141, 149, 199–201 (*see also* moon landing)
  Apollo 12, 133, 136, 141
  Apollo 13, 140–144, 147
  Apollo 14, 156
  Apollo 18, 120
  Apollo 19, 120
  Apollo 20, 134
  Apollo 1 accident, 45–46, 50, 55, 70, 80, 140, 149
  flights after Paine left NASA, 151–152
  Lunar Surface Experiment Package, 121
  Paine on naming of spacecraft, 85–86
  reasons for success of, 152–153
  schedule, 63
  spending on, 93–95
Apollo-Soyuz Test Project (ASTP), 118, 185–187
Armstrong, Carl, 14
Armstrong, Jan, 139
Armstrong, Neil, xii, xiii, 85, 89, 91, 92, 183, 202
  as associate administrator at NASA, 138–139
  celebrations of, 103, 104
  moon landing, 96–98, 101
  National Commission on Space and, 172
Atlantic Naval Submarine Base, 11–12
atomic bombs, 24–26
Atomic Energy Commission, 109–110, 125
Augustine, Norman R., 194–198, 206
Aviation Week and Space Technology, 180

Barnes, Richard J. H., 108
Baxter, Richard, 206
BBC Overseas Far Eastern broadcasts, 17
Beggs, James M., 169, 185
Belov, Evgeniy, 117
Bendix, 150
Bergen, William, 141
Bevill, Whitey, 16
Bisplinghoff, Raymond L., 78
Blagonravov, Anatoly Y., 81, 113, 114
Boeing Company, 119, 127, 150
Borch, Fred J., 156
Borman, Frank, 71, 73, 77, 79
Bottoms, Bud, 207
Bowman, Julian, 66
Boyd, Alan S., 48
Bradbury, Ray, 206
Brandt, Willy, 140
Brezhnev, Leonid, 116
Buckley, Thomas, 18
Buran, 187–188
Bureau of the Budget, 45, 124, 125, 136, 149
Bush, George H. W., 190, 191, 192
Butterfield, Alexander, 78
B-2 Spirit, 164, 165
B-70 Valkyrie, 40

Carter, Jimmy, 165, 167–168, 181, 185
Central Intelligence Agency (CIA), 64
Cernan, Gene, 87
Chafee, Roger, 45
Challenger shuttle, 182, 192–193
Chapin, Dwight L., 146
Chicago Tribune, 148
China, 42, 43, 108, 161
Christian, George, 56
civil rights movement, 93–95, 134
Clancy, Tom, 206
Clarke, Arthur C., 33, 123, 206
climate change, 195
Cold War, the, 31, 55, 112–113, 168, 181, 185–186
  end of, 196
  TEMPO and, 39
Coleman, Paul J., 172
Collier's, 131
Collins, Michael, 92, 96, 101, 103, 104, 202
Columbia, 85, 100–101, 203
Columbia space shuttle, 128
Columbus, Christopher, 131, 151, 186
command-service module (CSM), 62–64, 80, 85–86, 100, 142
Conrad, Charles, 97
Convair, 127
Cordiner, Ralph J., 38
Cortright, Edgar M., 123
Covert, Betty, 146, 150
Cunningham, Walt, 61

Debus, Kurt, 66
Disney, Walt, 131
Dobrynin, Anatoli, 117
Donahue, Thomas, 181
Dryden, Hugh L., 46–47, 113
DuBridge, Lee A., 75, 77, 81, 122, 124–125, 136
Duns' Business Month, 165

Eagle, xii, 85, 98, 101, 133
Edison, Thomas, 35, 134
"Effect of a Molten Lead Bismuth Eutectic Alloy on Steel, The," 35
Ehrlichman, John, 137
Eisele, Donn, 61
Eisenhower, Dwight, 51, 76, 113, 169, 201
Emme, Gene, 152, 200, 201
energy crisis, 157–159
European Launcher Development Organization, 110–111
European Space Agency, 107
European Space Research Organization, 110–111
Evans, L. J., 141

Farley, Clare, 146
Field, George B., 172
First Man, 97
Fitch, William H., 172
Flanigan, Peter, 137, 159
Fletcher, James C., 118, 119, 128, 175, 181, 200, 201
Ford, Gerald, 181, 185, 203
Franklin, Ben, 14
Franklin, George C., 97
Freetown, Massachusetts, 1–2
Fremantle Submarine Base, 13, 32
Friedman, Lou, 185, 204
Frutkin, Arnold W., 108, 117, 185
Fuqua, Don, 167

Gagarin, Yuri, 102, 136
Gemini Project, 44, 80, 96, 97, 113, 139–140
General Electric, 35–37, 38, 48, 49, 52, 76
  energy crisis and, 157–159
  in the Far East, 160–163
  offer for Paine to return to, 155–156
  promotion of Paine to senior vice president, 162–163
  Soviet Union and, 159–160
  TEMPO, 39–44, 49, 158, 207
Gilruth, Robert, 66, 69, 80, 91, 97, 141
Gimber, Stephen H., 9, 15–19
Glasnost, 189
Glenn, John, 139
Goddard, Robert, 78, 89, 171
Gorbachev, Mikhail, 182, 186
Gordon, Richard, 97
Gouldthorpe, Hubert W., 156
Graham, Billy, 104
Great Depression, 3
Great Society campaign, 55, 137
Griffin, Gerald D., 80–81
Grissom, Gus, 45
Gromyko, Andrey, 186–187
Guam, 19, 20, 23, 25, 29

Hage, George, 69
Haggerty, Patrick E., 76
Haise, Fred, 140, 142, 143
Haldeman, H. R., 78, 137
Handler, Philip, 117
Hansen, James, 97
Harr, Karl G., Jr., 169
Hartman, Arthur A., 186
Hartmann, William, 175
Hashimoto, Mochitsura, 25–26

Hatsutaka, 21
Hawk, Earle C., 13, 14, 15
Health, Education, and Welfare (HEW), Department of, 95
Heavy-lift launch vehicle, 127, 179, 189, 192
Helo66, 101
Herzfeld, Charles M., 172
High Frontier: Human Colonies in Space, The, 204
Hoffman, Bill, 21
Hollings, Ernest F., 170
Hollomon, J. Herbert, 36, 38, 48–49
Hornig, Donald, 71
Housing and Urban Development (HUD), Department of, 94–95
How to Build the Racing Catboat Lark, 4
Humphrey, Hubert, 75
Huston, Vincent, 71

India, 111
intercontinental ballistic missiles (ICBM), 40
International Astronautical Federation Congress, 118

Japan. *see* World War II
Jikitanchiki, 17
Johnson, Lady Bird, 59
Johnson, Lyndon B., 42, 47, 148, 181
  Great Society program, 55, 137
  National Air and Space Museum and, 201–202
  Soviet Union and, 113
  support for NASA, 50–51, 55–59, 69, 71–72, 74, 81, 99
Johnson, Richard G., 173, 175
Johnson Space Center (see Manned Spacecraft Center)
Johnson, U. Alexis, 150
Jones, James R., 71
Jones, Reginald, 155, 159, 162–163
Jones, Thomas V., 164

Keldysh, Mstislav, 115–117, 118
Kennedy, John F., xiv, xv, 40, 46, 47
  support for the space program, 55, 76, 113, 123, 140, 152, 168
  Ted Kennedy and, 89–91, 137
Kennedy, Robert, 73, 75
Kennedy Space Center, xii, 65, 84, 93-94
Kennedy, Ted, 89–91, 137
Kerrebrock, Jack L., 172
Keyworth, George A., 171, 173, 207
King, Martin Luther, Jr., 73, 93
Kirkpatrick, Jeane J., 172, 175
Kissinger, Henry, 107–108, 110, 114, 115
Komarov, Vladimir, 64
Korolev, Sergei, 64
Kraft, Chris, 66, 71, 74, 97
Kranz, Gene, xii

Laird, Melvin R., 78, 125
L'Amour, Louis, 182
Langley Research Center, 123
Ley, Willy, 131
Lindsay, John, 104
Lockheed Missiles and Space Company, 76, 127, 165

Lockwood, Charles A., 28
Long Beach Naval Shipyard, 32
Louisiana Purchase, 131
Love, John, 159
Lovell, James, 78, 97, 140–143
Low, George, 63–67, 71, 79–80, 97, 116
  Apollo 13 and, 143, 144
  Apollo 9 and 10 projects and, 86–87
  Paine's resignation and, 151
  Ronald Reagan and, 169
Low, Mary Ruth, 151
Lunar Excursion Module, 84
lunar module (LM), xi-xii, 63–64, 84–87, 92, 97, 142
Lunney, Glynn S., 118, 143
Luttenberger, Philip, 158
Lynn Instrumentation Laboratory, 36
Lynn, Vera, 17

MacArthur, Douglas, 20
Macy, John W., Jr., 47–51
Magna Carta of Outer Space, 206
Manned Spacecraft Center, xi-xiii, 55, 65, 69, 80, 81, 97, 118, 141
Manned Space Flight Network, 71
Mark, Hans, 91, 133, 147
Mars mission plans, 125–132, 173, 190–191, 194, 203–204
Mars Observer, 206–207
Mars Underground, 203
Marshall Space Flight Center, 65–66, 123, 129, 138, 206
Martin, Charles, 201
Martin Marietta, 127, 194
Massachusetts Bay Colony, 1
Matthew Fontaine Maury High School, 4
Mayflower, 2
Mayflower Compact, 2
Maynard, Owen, 63
McCall, Robert, 175
McCormack, John W., 99
McDivitt, Jim, 85
McDonnell Douglas, 127
McDowell, J. M., 28–30, 31
McVay, Charles, 26
Mendenhall, Bill, 14, 20
Mercury Project, 76, 80
Miller, George P., 99, 130
Millionshchikov, Mikhail Dmitrievich, 150
Mir space station, 187
Mission Operations Control Room, xi–xii, 141
Mississippi Test Facility, 120
moon landing, the, 61, 98–100, 133–134, 149
  celebrations of the success of, 103–106
  Columbia reentry after, 100–101
  global response to, 101–103
  metallurgy and, 179–180
  Paine on repeating on, 134–135
  plans for flag and plaque, 91–93
  televising of, 98
Mueller, George, 63–69, 70, 71, 80, 116, 143
  Apollo 11 and, 84, 86–87, 95, 97
  departure from NASA, 139–140
  space shuttle and, 119, 128

Murray, Bruce, 122
Myers, Dale D., 143

NASA, xii, xv. *see also* Apollo Project; moon landing; Space Race
  Apollo-Soyuz Test Project (ASTP), 118
  bad publicity, 192
  budgets and planning, 81–86, 119–120, 135, 149
  command-service module (CSM), 62–64, 80, 85–86, 100, 142
  Electronics Research Center, 120
  Gemini Project, 44, 80, 96, 97, 113, 139–140
  George Low and, 63, 64–66, 67, 71, 79–80
  under James E. Webb, 45–57
  under Jimmy Carter, 167–168
  John W. Macy, Jr. and, 47–51
  lunar module (LM), xi-xii, 63–64, 84–87, 92, 97, 142
  under Lyndon Johnson (*see* Johnson, Lyndon B.)
  as management test bed, 54
  management council, 138
  Mars mission, 125–132
  Mission Control, 53, 89, 98, 118, 141, 142
  Mission Operations Control Room, xi–xii
  National Commission on Space (NCOS) (see also *Pioneering the Space Frontier*), 171–180
  Office of the Administrator, 46, 52–54, 66
  Office of Manned Space Flight, 63, 140
  Paine appointed to, 48–56
  Paine's deputy administrator at, 78–81
  Paine's relationships with others at, 61–74
  Paine's resignation from, 144–147, 147–148
  plans for flag and plaque for moon landing, 91–93
  proposals for joint experiments with the Soviets, 115–118
  under Richard Nixon (*see* Nixon, Richard)
  under Ronald Reagan, 168–172
  Space Exploration Initiative (SEI), 190–198
  space shuttle, 125, 126–129, 192–193
  Ted Kennedy's criticism of, 89–91, 137
NASA One, 141
National Academy of Sciences, 117, 181, 206
National Aeronautics and Space Act of 1958, 181
National Aeronautics and Space Council (NASC), 181–182
National Air and Space Museum, 172, 201–203
National Commission on Space, 171–172, 195
  meetings, plans, and findings, 172–180
  reactions to, 180–183, 189

National Space Club, 171
Nelson, Bill, 186
Nelson, Horatio, 70
NERVA nuclear engine, 129, 147
Newell, Homer E., Jr., 53, 57, 77, 144
New York Times, 133
Nimitz, Chester W., 18
Nixon, Pat, 101, 104, 133
Nixon, Richard, xii, 71–72, 73, 75–78, 81, 104, 148, 181, 185, 200
  address to the United Nations, 107
  Apollo 13 flight and, 142–143
  Columbia splashdown and, 100–101
  Department of Energy, 158
  global travel, 104, 107–108
  lukewarm support of the space program, 107, 124, 126, 130, 132, 135–136, 137
  moon landing and, 92–93, 98
  National Air and Space Museum and, 202
  negotiations with the Soviet Union, 112, 114–115
  praise for GE, 160
  resignation of Paine and, 146–147
  Space Task Group, 124–125, 130
  Watergate scandal, 159, 164
North American Aviation, 46, 63
North American Rockwell, 127, 141
Northrop, Jack, 164
Northrop Corporation, 163–166, 203
nuclear power, 157–158

Office of Management and Budget (see Bureau of the Budget)
Office of Naval Research, 34
Ogilvie, Richard B., 104
O'Neill, Gerard K., 172, 204–205
OPEC oil embargo, 157
Operation Downfall, 19
Operation Magic, 27
Orwell, George, 205
Otten, Ada Louise, 3, 6
Outer Space Treaty, 112

Page, Hilliard W., 76
Pain, Ralph, 1, 2
Paine, Barbara, 199, 206
  background of, 32
  marriage to Tom Paine, 20, 32, 33, 37, 51, 52, 108, 151
  visit to the Soviet Union, 188, 189
Paine, Frank Eugene, 2
Paine, George Thomas, 2–4, 6
Paine, Janet Augusta, 3
Paine, Jemima, 2
Paine, Job, 2
Paine, Thomas, 2
Paine, Tom (Thomas Otten)
  1970s cuts to NASA and, 135–137
  advocacy for a Mars program, 129–132
  ancestors of, 1–2
  on the Apollo 8 program, 73–74, 83–84
  applications to graduate schools, 33
  appointed to NASA, 48–56
  assignment to submarine duty, 11
  birth of, 3
  at Brown University, 5–6, 9

Paine, Tom (Thomas Otten) *(continued)*
celebrations of the moon landing and, 103–106
childhood of, 3–4
children of, 37, 52, 78, 156
civil rights movement and, 93–95
on *Columbia*'s splashdown, 100
Commendation Ribbon received by, 18–19
on commercial space transportation, 174, 179–180
death of, 206
decision to leave the Navy, 31
on defending the NASA budget, 81–84
duties after the end of World War II, 23–31
"Energy Czar" and, 159
enlistment in the U.S. Navy, 9–10
failure of third landing attempt and, 143
first visit to the Soviet Union, 159–160
on future missions to the moon, 134–135
on future technology and space, 204–205
at General Electric, 35–37, 38–44
as global emissary, 108–114, 188
as head of NASA, xii, xv, 58–59
importance of space program to, xvi, 120–122, 124
interest in sailing and seafaring, xiv, 4–5, 10
interest in space exploration, xiii
on James Webb's resignation, 57–58
on mining the moon,179–180
job offers after leaving NASA, 155–156
last day at NASA, 150–151
legacy of, 206–207
marriage of, 32
in the Mission Operations Control Room, xi–xii
plans to attend MIT, 5, 6
on Mars governance, 204
Mars flag designed by, 207
move to Washington, D. C., 52
NASA budgets and planning and, 81–86
National Air and Space Museum, 201–203
National Commission on Space and, 171–180
on creating a National Space Exploration Agency (NSEA), 194
as "Navy brat," 3
negotiations with the Soviets, 113–118
at the Northrop Corporation, 163–166
parents of, 2–4
plans for NASAs future, 137–140
politics and, 51
relationships with others at NASA, 61–74
resignation from NASA, 144–147, 147–148

return to General Electric, 157–159
Richard Nixon and, 75–78
Ronald Reagan and, 169–170
second visit to the Soviet Union, 186–188
as senior vice president at GE, 162–163
Space Exploration Initiative (SEI) and, 190–197
on the Space Race, xiv–xv
space shuttle program and, 126–127, 129, 192–193
speech to the United Nations, 104–106
at Stanford University, 33–35
third visit to the Soviet Union, 188–190
Thomas Paine Associates, 166
views on the scientist versus the engineer, 121
wartime journal, 15, 21, 29
at the Washington National Cathedral, 199–201
on why Apollo worked, 151–153
Pan American Airways, 175
Paris Peace Accords, 160
Pasteur, Louis, 131
Pearl Harbor attack, 9, 12, 30
Pearse, Barbara Helen Taunton. *see* Paine, Barbara
Pearse, Henry William Taunton, 32
Pearse, Marguerite Jones, 32
Perez-Marin, Antonio, 150
Petrone, Rocco, 71, 143
Phelps, Ben, 20–21
Phillippe, Gerald L., 41
Phillips, Sam, 68, 69, 78–79, 86, 97
Pioneering the Space Frontier: An Exciting Vision of Our Next Fifty Years in Space, 175, 180, 182–183, 189, 191
Planetary Society, 122, 203–204
Plymouth Colony, 1–2
Pratt & Whitney, 36
Press, Frank, 167
Princi, M. A., 36

Quayle, Dan, 129, 190–194, 196–197, 206

Ramo, Simon, 76
Raymond, Richard C., 40
Reagan, Ronald, xvi, 35, 104, 167, 168–172, 175, 182, 185–186
Rees, Eberhard, 66, 70
Rickover, Hyman G., 34
Rogers, William, 114
Root, L. Eugene, 76
R-14 submarine, 11

Sarabhai, Vikram A., 111
Saturn IB rocket, 62, 116
Saturn V rocket, 62, 64, 66–67, 71, 120, 133, 149
Sayre, Francis B., Jr., 199–201
Scheer, Julian, 68, 90, 93, 103, 146
Schirra, Wally, 61
Schmidt, Ed, 41, 42, 43
Schneider, William, 66
Schriever, Bernard A., 76–77, 172

Schweickart, Rusty, 85
S-class submarines, 11–12
Scott, Dave, 85
S-75 Dvina, 40
Seaborg, Glenn T., 109
Seamans, Robert C., Jr., 46–47, 72, 78, 85, 124–125, 127
Shapley, Willis, 68, 91, 92, 146
Shepard, O. Cutler, 34, 35, 37
Sherrod, Robert, 42, 43, 72, 155
Sino-Japanese War, 12
Skylab, 80, 116, 120, 123–124, 129, 134
Slayton, Deke, 66, 97, 186
Smith, Marcia S., 172, 173
Smithsonian Institution, 200
social activism/justice, 55, 134, 148
Soule, George, 2
Southern Christian Leadership Conference, 93
Soviet Union, the, 31, 43, 64, 73, 99–100, 102, 108, 121–122, 185
  declining space program efforts, 135–136
  détente with, 112
  fall of, 196
  Glasnost, 189
  Paine and negotiations with, 113–118
  Paine's first visit to, 159–160
  Paine's second visit to, 186–188
  Paine's third visit to, 188–190
  Richard Nixon and, 112, 114–115
  Ronald Reagan and, 168, 182, 185–186
  space program in the 1980s, 187–188
  Sputnik, xiv–xv, 37, 98, 113, 136, 169, 186
  as U.S. top priority in international relations, 112–113
Spacecraft Tracking and Data Acquisition Network, 113
Space Exploration Initiative (SEI), 190–198
Space Race, xiv–xv, 37, 98–100, 102, 112–113, 181
space shuttle/space station, 123, 147, 170, 171, 177, 185, 187, 192–193
space station *Freedom,* 177, 193, 197
Space Task Group, 124–125, 130
Sputnik, xiv, 37, 98, 113, 136, 169, 186
Stafford, Tom, 87, 186
Stand By to Surface, 206
Stanford University, 33–35
"Star Wars," 170
Steele, Hoyt P., 160–161
Stevens, James E., 27
Stolpman, Paul, 16
Strategic Defense Initiative, 170, 171, 182, 186–187, 206
submarines, xiv, 10–13
  nuclear, 37
Sullivan, Kathryn D., 172
Swigert, Jack, 140–143
synfuels, 158–159

Taiwan Power Company, 161
Teague, Olin E., 116, 167
TEMPO, 39–44, 49, 158, 207
Texas Instruments, 76

Thomas Paine Associates, 166
Time Life, 42, 155
Tindall, Howard W., Jr., 86
Townes, Charles H., 77, 80
Trans-Alaska Pipeline System, 157
Treaty on the Non-Proliferation of
Nuclear Weapons, 112
Truly, Richard H., 190, 194
Truman, Harry, 45, 201
Truszynski, Gerald, 71
TRW, 76
2001: A Space Odyssey, 123

U-boats, 11, 15, 206
United Nations, 91, 104
  Outer Space Committee, 113
U.S. Air Force, 37
U.S. Naval Academy, xiv, 5, 10, 206
U.S. Navy, 2–3, 9–10, 26, 31
  nuclear submarines, 37
  Office of Naval Research, 34
  personnel lost in World War II, 20–21
  submarines, xiv, 10–13
U.S. State Department, 43, 68, 111,
125, 160
USS *Arizona,* 30
USS *Barb,* 21
USS *Billfish,* 13
USS *Boarfish,* 31
USS *Enterprise,* 37
USS *Euryale,* 23
USS *Herndon,* 19
USS *Herring,* 21
USS *Hornet,* 100, 101
USS *Indianapolis,* 26
USS *Lagarto,* 20–21
USS *Missouri,* 20
USS *Pelias,* 13
USS *Pompon,* 14, 15–21, 23, 31, 32
USS *Tuna,* 13

Van Allen, James, 122
Vietnam War, 50–51, 55, 72, 73, 82,
134, 135, 160
Viking 1, 203
Viking 2, 203
Viking Mars mission, 120, 203
Voice of America, 111
von Braun, Wernher, 65, 70, 104, 129,
131, 137–138, 147

Wahlin, Walter H. F., 14–15
Wall, Frank, 14
Washington National Cathedral,
199–201
Washington Star, 77, 144
Watergate scandal, 159, 164
Webb, David C., 172
Webb, James, 45–57, 63, 68–70, 77–78,
99, 102, 112, 123
Webster, Daniel, 131
White, Ed, 45
Wilkening, Laurel L., 171, 173, 175,
191, 195
Williams, Hosea, 93
Wilson, Woodrow, 199
Winfield, Rodney M., 199
World War II, 7, 9, 83
  activities after the end of, 23–31
  atomic bombs, 24–26

World War II *(continued)*
Operation Downfall, 19
Paines' service in, 12–31
Pearl Harbor attack and, 9, 12, 30

Xerox Corporation, 155

Yeager, Charles E., 172
Yemen, 43
Yen Chia Kan, 161
Young, John, 87

Ziegler, Ron, 146